e-Stocks

FINDING THE HIDDEN BLUE CHIPS
AMONG THE INTERNET IMPOSTORS

PETER S. COHAN

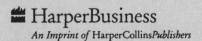

HarperBusiness

An Imprint of HarperCollins*Publishers*

HarperCollins books may be purchased for educational, business, or sales promotional use. For information please write: Special Markets Department, HarperCollins Publishers, Inc., 10 East 53rd Street, New York, NY 10022.

FIRST EDITION

Designed by Michael Mendelsohn at MM Design 2000, Inc.

Library of Congress Cataloging-in-Publication Data
Cohan, Peter S., 1957–
    E-stocks : finding the hidden blue chips among the Internet impostors / Peter S. Cohan.—1st ed.
        p.   cm.
    Includes bibliographical references and index.
    ISBN: 0-06-662083-X
    1. Internet industry—Finance.   2. Investments.   I. Title.

HD9696.8.A2.C635 2001
332.63'22'02854678—dc21
                                                                00–053905

    01   02   03   04   05   RRD   10   9   8   7   6   5   4   3   2   1

To Sarah, Adam,

and children everywhere.

_____

They are why investment matters.

CONTENTS

ACKNOWLEDGMENTS

THIS BOOK COULD NOT HAVE BEEN WRITTEN without the help of many people. I am grateful to all the professionals whose insights helped shape my thinking for this book, including Rob Burgess (Macromedia), Charlie Giancarlo (Cisco Systems), Ben Gordon (3Plex.com), Marco Iansiti (Harvard Business School), Elon Kohlberg (Harvard Business School), Tim Koogle (Yahoo), Baruch Lev (New York University), Jonathan Roosevelt (Epesi Technologies), Asif Satchu (formerly of Ariba), Tanya Yannas (3Plex.com), Mickey Butts (*The Industry Standard*), Jeff Davis (*Business 2.0*), Bambi Francisco (*CBS MarketWatch*), Luke Mitchell (*The Industry Standard*), Beth Piskora (*New York Post*), Amey Stone (*Business Week Online*), Mark Veverka (*Barron's*), and Jack Willoughby (*Barron's*).

I also appreciate the help of my colleagues Geoff Fenwick (The Balanced Scorecard Collaborative), Eric Stang (Lexar Media), Fred Hajjar (Accenture), Peter Laino (Monitor Clipper Partners), Jake Wesner (Perot Investments), and Mordechai Fester (Syndeo Corporation). I am very grateful for the support of Adrian Zackheim and David Conti of HarperCollins and to my agent, David Gernert (The Gernert Company), whose persistence made this possible.

Finally, I would like to offer special thanks to my wife, Robin, who patiently tolerated the seemingly endless weekends of writing and revisions.

PETER S. COHAN
Marlborough, MA
January 2001

INTRODUCTION

INVESTING MONEY is the single most important business activity in our economy. If done well, investing can create enormous wealth that can provide a secure future for generations. (While living off inherited wealth can create problems, many of the most successful investors have created estates that give 95% of their fortunes to charity and establish trusts to make sure that estate beneficiaries do not become debilitated by their inheritances.) If investment is done poorly, it is easy to wipe out a family's life savings in a short period of time and turn a planned comfortable retirement into a nightmare of impoverished old age.

For some, the concept of Internet stocks and investing simply do not go together. Many people purchased certain e-commerce stocks at the peak of the market frenzy of early 1999 and subsequently saw those stocks collapse in value during the April 2000 Internet stock meltdown. These individuals may never want to hear about investing in Internet stocks again. Others had the good fortune to invest in privately held Internet companies and to enjoy returns in excess of 2,000% on their investment in 12 to 18 months following an initial public offering (IPO) or an acquisition. These examples suggest that investing in Internet stocks has the potential to generate big wins *and* enormous losses.

The world of investing in Internet stocks has changed completely since April 2000; however, the structure of the product/service markets in which Internet companies compete has remained relatively stable. It is this distinction between the market for securities and the performance of companies and industries which these securities theoretically represent that underlies this book.

Prior to April 2000, a rare confluence of factors enabled many companies, whose inherent profit potential was very weak, to obtain financing and go public. As we will explore in this book, these forces included **public investors** with an unlimited appetite for companies with ".com" appended to their names; **venture capital firms** eager to finance dot-coms as a result of returns often exceeding 15 to 20 times their investment in 18 months; **investment banks** happy to earn their 7% fees for underwriting the IPOs of these prematurely born companies; and **the media,** which amplified the optimistic chorus led by these banks' analysts.

Following April 2000, the capital markets turned their backs on these companies, and many of them collapsed. The turning point was a growing realization that many of the dot-coms were running out of cash. Investors decided that they did not want to be left holding an empty bag, so they sold—destroying roughly 90% of the market capitalization. After the companies were abandoned by public investors, the venture capital firms, underwriters, and press followed suit. With little in the way of revenues and burgeoning operating losses, about 360 of these dot-coms disgorged 31,000 employees between December 1999 and November 2000.

Despite all this upheaval in the capital markets, the structure of the product/service markets in which Internet companies compete has remained relatively stable. My book *Net Profit,* which was written in 1998 and published in April 1999, analyzed the structure of the product/markets in which the dot-coms competed and concluded that the vast majority of them would not be able to earn profits over the long term and would therefore collapse.

The point of this citation is simple. In the short run, capital markets over- and undershoot the financial performance of the companies whose stocks they price each day. In the long run, however, the capital and product markets tend to coverage. In this distinction lies an opportunity for investors to profit.

This book is about helping investors to avoid disastrous Internet investments and to exploit spectacular ones. Through investing and in conducting the research for this book I have come to a surprising conclusion: There is indeed a method that investors can follow to help increase the odds of making money by investing in Internet stocks. The reason that investors should read this book is to understand this method well enough to use it on their own behalf. To explain this method, *e-Stocks: Finding the Hidden Blue Chips Among the Internet Impostors* addresses the following questions:

- What is the value of the Internet as a source of new businesses?
- What is an Internet stock? What are the various segments into which these stocks can be categorized?
- How have Internet stocks performed relative to the broader market indices?
- What factors have driven the relative performance of these Internet stocks?
- Which Internet stocks have generated the highest returns for investors? Why?
- Which Internet stocks have performed the worst? Why?
- Which Internet stocks have been the most volatile? Why?
- What quantifiable factors are the most effective predictors of the relative performance of Internet stocks?
- What are the characteristics of the Internet companies that have provided the highest returns for private-equity investors?
- What are the implications of all these findings for investors?

WHO SHOULD READ THIS BOOK

This book is written for anyone who wants to achieve significant financial goals such as paying for children's education, securing retirement, or leaving a significant inheritance for children and grandchildren. In that sense, the book is written for almost everybody. Readers who are likely to make the most use of this book are those who want to participate in the exceptionally high potential returns of investing in Internet stocks while avoiding some of the potentially disastrous pitfalls. Readers should understand, as we will discuss in Chapter 13, that Internet stocks should constitute only a portion of the portfolio of any investor. Because of the risks of Internet stocks, it is dangerous to depend on them exclusively as a means of achieving long-term financial goals.

RESEARCH

The book is based on research into the performance of roughly 200 publicly traded Internet companies. Quantitative analysis has been used to pinpoint the variables that do the best job of explaining the variation in the performance of these Internet stocks in the aggregate and by Internet business segment.

This quantitative research has been supplemented by asking industry executives and stock analysts for their opinions about what is driving the performance of Internet stocks. This qualitative research included questions such as:

- How do Internet stocks perform relative to the broader market indices?
- What factors are most critical in driving the relative performance of Internet stocks?
- What correlation exists between the relative stock market performance of Internet companies and their performance in their product/ service markets?
- What methods determine which Internet stocks to invest in and which to sell?
- How can an investor decide when to purchase and when to sell Internet stocks?

This book is also based on a significant amount of data collected on the performance of Internet industries and individual companies. The data on industries help investors assess the size, growth rate, and key profit drivers for the average industry participant. The data on individual companies help investors understand the profitability and profit potential, the effectiveness of a firm's competitive strategy, the quality of the firm's management, the extent to which the firm's management makes optimal use of its financial resources, and the firm's relative value from the perspective of the stock market.

ORGANIZATION OF THIS BOOK

This book is divided into three parts. The first part, Chapters 1–3, presents the conceptual framework on which the book is developed by focusing on an analysis of the value of the Internet as a generator of new business and by evaluating the performance of the Internet index relative to other indices. The second part, Chapters 4–12, evaluates the performance of specific stocks within nine Internet business segments and pinpoints the drivers of their relative performance. Chapter 13, the third part, concludes by describing the implications of the findings in the first two parts for investors.

Chapter 1 develops a framework to help explain the revenue growth rates of Internet companies, which in turn help drive Internet stock valuations. Chapter 2 examines the factors that determine the value of different types of networks. It distinguishes between one-to-many networks (in which sellers offer products or services to many buyers) and many-to-many networks (where all network participants can be buyers and sellers). For each type of network, Chapter 2 presents the results of a simulation model that quantifies the potential value of transactions taking place over the network. The chapter quantifies the relative impact of various factors—such as growth in the number of network participants, increased probability of transactions, lower prices, product/service profit margins, transaction frequency—on each type of network's value. Chapter 2 concludes by describing general principles for enhancing the value of a network and articulating the linkage between these general principles and the market performance of Internet stocks both in the aggregate and by Internet business segment.

Chapter 3 evaluates the performance of Internet stocks in general relative to broader market indices. It analyzes the performance of the Internet index in periods of increasing and declining stock prices in comparison to the prices of the Dow Jones Industrial and NASDAQ indices. It pinpoints the quantitative factors—interest rates, earnings growth rates, money flows, etc.—that are most tightly correlated with variation in the value of these indices. Chapter 3 also analyzes the key drivers that explain the differences in relative performance among the Internet, DJIA, and NASDAQ indices during these up- and down-market periods. The chapter presents the results of interviews with leading analysts and industry executives to offer their explanations for the key factors driving the relative performance of the Internet stocks. It develops some general principles of investing in Internet stocks. Chapter 3 concludes by introducing the concept of the nine Internet business segments from *Net Profit*.

Chapter 4 analyzes the stock market performance of a sample of Internet infrastructure stocks. It continues by applying the Internet Investment Dashboard (IID) analysis to the Internet infrastructure segment. It analyzes the size, growth rate, and inherent economic bargaining power of Internet infrastructure firms. It then assesses the importance and relative strength of the closed-loop solutions provided by these Internet infrastructure firms. Chapter 4 continues by analyzing the management integrity and adaptability of Internet infrastructure firms. It assesses the relative strength of their brand families and financial effectiveness. It concludes the IID analysis by assessing the market valuation of the Internet infrastructure firms. Chapter 4 concludes by detailing six principles that investors can follow to identify winning Internet Infrastructure stocks and avoid weaker performers.

Chapters 5 through 12 follow the same kind of analysis for the other eight Internet business segments. In particular, the chapters accomplish this with specific examples from companies, including:

- Web consulting (Chapter 5) uses examples from companies such as Sapient, Scient, Viant, and DiamondCluster International;
- Internet venture capital (Chapter 6) uses examples from Internet Capital Group and CMGI;
- E-commerce (Chapter 7) uses examples from Amazon.com, Vertical-Net, Ariba, and Commerce One;
- Web portals (Chapter 8) uses examples from Yahoo, Terra Lycos, Excite@Home, and GoTo.com;
- Web security (Chapter 9) uses examples from Check Point Software, VeriSign, ISS Group, and Network Associates;
- Web content (Chapter 10) uses examples from CNet, Forrester Research, and TheStreet.com;
- Internet service providers (Chapter 11) uses examples from AOL, EarthLink, Exodus Communications, and IDT Corporation; and
- Web tools (Chapter 12) uses examples from Spyglass, BroadVision, Macromedia, and Vignette;

Chapter 13 shows investors how to use the findings in the foregoing chapters. The chapter begins by describing factors that investors can use to evaluate whether they should be investing in Internet stocks at all. It then helps investors decide whether they should invest in an index of Internet stocks or specific Internet business segments. Chapter 13 continues by presenting the criteria that investors should use to pick specific Internet companies in which to invest. It describes criteria outlining when to buy and when to sell the stocks. The chapter presents a process for monitoring an existing portfolio of Internet stocks. Chapter 13 concludes by showing how to apply early-warning indicators of waxing and/or waning Internet investment performance to choosing when to sell specific stocks in the portfolio or when to add to existing Internet investment positions.

Let's enter together the exciting world of Internet investing!

Chapter 1

E-Stocks:

Finding the Hidden Blue Chips Among the Internet Impostors

Consider two Internet stocks—Check Point Software, up 2,571% since August 1998, and iVillage, down 99% from its all-time high in April 1999. Check Point Software is an Internet security software company whose stock price rose almost in a straight line from a split-adjusted $5 in August 1998 to $133.56 in December 2000. By contrast, iVillage is a Web site for women whose stock fared less well. According to MSNBC, iVillage's 3.65-million-share initial public offering was underwritten by Goldman Sachs in March 1999 at an offering price of $24 per share. At its open, iVillage traded up to $96 a share and climbed briefly to $120 in the weeks following (Byron, 2000). In December 2000, iVillage traded at $1.06 per share, a decline of 99% from its all-time high, of 98% from its opening trade, and of 96% from its official opening price.

The contrast in stock market performance between these two firms highlights a number of important issues. Is there any link between the underlying performance of the business as measured by its reported financial statements and its stock market performance? What role does the inherent profit potential of the industry play in determining the stock market success of an Internet company? How important is relative market share in determining the performance of an Internet stock? What is the difference between a well-managed Internet company and a poorly managed one, and how do these differences affect the relative stock market performance? What impact do changes in supply and demand for the stock itself have on the firm's relative stock market performance?

Check Point Software as a company has benefited from its market leadership in a structurally attractive market. As we will explore in much greater depth in Chapter 9, Check Point Software is the leader in firewall software. Firewalls protect an organization's Web site from intrusion by hackers. The industry is attractive because it is big, growing quickly, and is able to charge a high price for quality products. Check Point Software has used its firewall leadership to knit together an entire suite of security products that give corporate network managers a higher level of network protection. Check Point Software also benefited from excellent financial results in 2000, including 88% revenue growth and 115% net income growth in the nine months ending September 2000.

iVillage has been damaged by its participation in an industry segment that is structurally unattractive and full of aggressive competitors. The company's strategy is to create Web content designed to draw women visitors whom Web advertisers will pay to reach. For the nine months ended 9/30/00, iVillage revenues reached $57.7 million, up from $19.9 million. iVillage's net loss for the period grew to $169.7 million, up from $79.7 million. iVillage gets 75% of its revenues from unprofitable Web advertising and the balance from an unprofitable online baby-products operation. In addition, *The Delaney Report* notes that iVillage has had a large amount of management turnover (*Delaney Report*, 1999). In a nutshell, iVillage is a weak company in a lousy industry, and its stock price, which declined 95% to $1.06 a share during the 52 weeks ending December 29, 2000, reflects the market's understanding of that fact.

While these two cases are by no means the most extreme examples of Internet investment wins and losses, they do illustrate an important point: Investing in Internet stocks is a game in which people can make and lose huge amounts of money. In the last several years, Internet investing has become huge. For example, as the financial weekly *Barron's* pointed out, Internet.com's Internet Stock List of publicly traded firms that receive more than half their revenues because of the Internet counted about 270 firms in February 2000, with total market capitalization in excess of $1.5 trillion. Those in the list that were public in 1997 had a combined market capitalization of $50 billion (Donlan, 2000). As we noted in the introduction, by December 2000, a significant portion of that value had been destroyed.

To help investors win the game of investing in Internet stocks, *e-Stocks: Finding the Hidden Blue Chips Among the Internet Impostors* develops seven key concepts:

- Securities markets as five competing belief systems
- Six rings of securities market information
- Structural evolution of networked business
- Nine Internet business segments
- Four tests of a winning Internet company
- Internet Investment Dashboard
- The seven-step Internet investment portfolio-management process

While the remainder of this book will develop each of these concepts in much greater detail, we will introduce the concepts here and show how they can help investors win the game of Internet investing.

SECURITIES MARKETS AS FIVE COMPETING BELIEF SYSTEMS

An important reason that securities markets function is that *different groups of investors* with *varying investment objectives* find themselves on opposite sides of a securities transaction because of their *differing belief systems*. Simply put, a buyer who believes a stock's price will rise purchases that stock from a seller who believes its price will fall.

When this book advocates a particular belief system, we will introduce that belief system by putting it in the context of competing systems. By doing so, it will become clearer why the belief system advocated in this book is likely to produce superior results for investors in Internet stocks.

As noted, securities markets depend on different groups of investors. One of the most significant trends over the last decade has been that the balance of power in the securities markets has tilted in favor of the individual. According to the *Federal Reserve Bank Survey of Consumer Finance*, released in May 1998, the proportion of American households that own stocks, either directly or through mutual funds and pension plans, has more than doubled, from 19% in 1983 to 41% in 1995, the latest year for which official numbers were available (Federal Reserve, 1999). Morgan Stanley's merger with Dean Witter was an acknowledgment of the importance of this shift. While Morgan Stanley was a leader in institutional banking services, it was relatively weak in the growing area of providing brokerage services to individuals, an area where Dean Witter was strong.

The point of these examples is that there are different groups of investors in the market and that the relative level of importance of these groups is changing. The reason that these differences are important is that the different groups have different investment objectives. For example, most households are saving for major life events like the purchase of a house, paying for children's education, and retirement. Institutional investment managers purchase stock to have the highest quarterly returns for their funds relative to a specific benchmark.

Investment behavior is guided by one of five common belief systems. Before detailing the five, let's define a "belief system." A belief system is a set of rules that help people feel more comfortable making a decision despite the uncertainty of the future outcome of that decision. Simply put, people know that the future is unknowable, yet they recognize that if they do not act, they have no chance of gaining. As a result, people develop belief systems that help them act despite the unknowable outcome of their actions.

While belief systems differ in their details, they share common structures. For example, belief systems have rules that create a linkage between a "measurable" system attribute (such as the inherent value of a company) and a decision (such as whether to buy or sell stock in that company). Belief systems depend for their power on a specific relationship between a leader (e.g., Warren Buffett) and a group of followers (such as "value investors"). Belief systems coexist in a competitive system that causes the belief systems to wax and wane in their power to predict investment behavior. In short, belief systems evolve. For example, while Warren Buffett's investment vehicle, Berkshire Hathaway, had excellent investment performance through much of the 1980s and 1990s, its 1999 performance was so disappointing that Buffett may be losing some of his followers.

The securities markets host five competing belief systems. Value investing is one such belief system. Introduced by Benjamin Graham's *The Intelligent Investor* in 1945, the basic notion of value investing is that a company has an inherent value. This inherent value is some combination of the liquidation value of its balance sheet and its earnings potential. If the business is publicly traded, a hardworking investor should be able to estimate the inherent value of that business and compare its inherent value to its stock market capitalization. If the inherent value is below the market value, the investor should sell the stock. If the relationship is reversed, the investor should buy.

A second belief system, which has been popularized recently by so-called day traders, is called momentum investing. Momentum investing involves a fairly simple concept: Understanding the business is irrelevant; what matters is whether the stock price is rising faster than other firms' stock prices. The momentum investor believes that what is going up will go up some more, thereby driving the momentum investor to buy. If a stock is not going up, it should be shunned. And if a stock is dropping rapidly in price, the momentum investor believes that there is money to be made by selling the stock short. The momentum investing philosophy is quite popular among day traders, who sell all their stock at the end of each day and therefore can only make money on minor fluctuations in stock prices during daily trading hours.

A third belief system is called technical analystis. Technical analysts make investment decisions based on patterns in stock prices over time. For example, technical analysts will chart a stock's daily prices against a "moving average" of the stock. This moving average, for example, might add up the stock prices for the previous 30 days and then divide that number by 30. Each day, that average would be recalculated by dropping the price from 31 days ago and adding the most recent price. Technical analysts chart the moving average and the daily price of a stock on the same graph and then make decisions about whether to invest based on the relationships between the two graphs. For example, if the daily price graph crosses above the moving average for the first time in a long time, technical analysts will buy. Conversely, if the daily price drops below the moving average, the technical analysts will sell. While technical analysis still has its adherents, it had an extremely influential advocate in the early 1980s, when Joseph Granville made predictions about the direction of the overall market and caused major moves in reaction to his market calls. Granville used a version of technical analysis called on-balance volume, which he used to make his prophetic market calls. Eventually, Granville's predictions proved incorrect, and he was subsequently ignored.

A fourth belief system, one that is growing increasingly common, dispenses with any claim to a methodology and focuses simply on following the recommendations of specific market gurus. In early 2000, some of the more prominent of these gurus were Abby Joseph Cohen, a Goldman

Sachs market strategist; Mary Meeker, a Morgan Stanley Internet analyst; Henry Blodgett, Merrill Lynch's Internet analyst; and George Gilder, an independent technology guru frequently associated with *Forbes*.

Because of the strength of some well-publicized recommendations that turned out well, these gurus can move markets. For example, according to Dow Jones Newswire, on February 17, 2000, George Gilder mentioned Terayon Communication Systems and Xcelera.com in the online version of his newsletter. That same day Terayon gained 67³/₁₆, or 49.1%, to close at 204¹/₁₆, then closed on February 18 at 230, up 12.7%. Xcelera.com rose 59.3%, from 76½ to 205½, an all-time high, on February 17 and jumped 16.8%, or 34½, to close at 240 on the next day. Gilder established his following, in part, by identifying Qualcomm, a wireless-communications-equipment manufacturer, as a winning company before it became well known by the markets. Qualcomm stock increased 2,618% during 1999 (Byrt, 2000). But gurus are fallible. By December 2000, Terayon had plummeted to $4 amid order cancellations, and Xcelera had plunged to $3.25.

A fifth belief system—and the one that we will examine in *e-Stocks*—is that investors in publicly traded Internet stocks are venture capitalists. The techniques that the more independent-minded venture capitalists use to invest in startup companies can be particularly useful for picking Internet stocks in which to invest. We will explore the details of this approach later. Suffice it to say for now that the basic concept of the venture capital approach is to identify the company that is likely to become the leader in an attractive market. The venture capitalist tries to invest as much as possible in that company and shun the rest.

Each of the five belief systems has strengths and weaknesses. Value investing proved a very compelling approach for investors and still serves as the primary counterweight to the recent activity in Internet stocks. Value investing is the intellectual underpinning of the perennial question "How can you justify the high valuations of these Internet stocks?" The problem with value investing is that it has no explanatory power for the $1.5 trillion in market capitalization we mentioned earlier.

Momentum investing seems to have gripped the minds of day traders and probably more institutions than we would care to know. The problem with momentum investing is that it has the greatest potential to lead to a collapse in prices, along the lines of the Dutch Tulip Bulb frenzy of the 17th century and many other popular delusions.

Technical analysis seems to have a scientific rigor in its methodology that does not appear based on much in the way of research. Nor are there any current adherents of technical analysis who seem to have the ability to move markets.

Guru investing is particularly dangerous because it involves placing too much trust in someone other than yourself. Since guru investing essentially involves surrendering one's own critical analysis, it is among the most dangerous. To the extent that one can anticipate what a guru will recommend and the duration of the guru's influence, guru investing can be a profitable strategy. Unfortunately, this extent is very limited.

e-Stocks: Finding the Hidden Blue Chips Among the Internet Impostors offers an approach to investing that is quite useful for investing in Internet stocks. As we will see, this approach has a set of indicators that are intended to help the investor understand important turning points in industry evolution and competitor strategies. As a consequence, this venture capital approach to investing is likely to be more robust and to produce better investment results. Only time will tell how accurate this prediction proves to be.

Before concluding this discussion of the five belief systems, it is important to explain how the different belief systems actually interact in the behavior of stock prices. One of the more common phenomena of Internet stock behavior is the upward price spiral. Such an upward price spiral can bring into play a number of the different belief systems. Here is an example of a situation in which two of the belief systems interact.

Let's take an example of a major e-commerce company, Amazon.com. On a particular day, let's say that Amazon.com's stock is trading at $100 a share. A value investor looks at the stock and concludes that since Amazon.com lacks any possibility of current earnings and that its physical assets are of minimal worth, it makes sense to sell Amazon.com short. The investor borrows shares from a broker and sells them at $100 a share, hoping to buy the shares back in the future at a lower price to repay the broker from whom he borrowed the shares. Spurred by a large buy order on the stock from another source, the price then rises to $101.00. This $1-per-share rise in price draws in the momentum investor, who goes into the market and purchases large quantities of Amazon.com to take advantage of the apparent upward momentum in the stock. As a result, Amazon.com rises in price to $103.00.

This rise in price actually forces some of the value investors who shorted the stock to cover their short positions owing to their broker's margin requirements. As a consequence, the value investor must go into the market to purchase a certain number of shares at $103 to return them to the broker. This particular purchase forces the price up even higher, to $104, which draws in more momentum buying. The result is that the price rises to $106, forcing more shorts to cover their position, causing an upward price spiral that is finally relieved at the end of the day when the momentum investors liquidate their Amazon.com holdings.

Momentum investing works well in reverse, too. The April 2000 crash in Internet stocks led investors to flee Amazon's stock in a downward price spiral where selling begat a lower price, which led to more selling. Between December 1999 and December 2000, Amazon's stock lost 80% of its value.

This interaction between competing belief systems is naturally based on different understandings of what beacons are of most significance to navigators on the uncertain waters of investing in Internet stocks.

SIX RINGS OF SECURITIES MARKET INFORMATION

The five belief systems that we just explored use different kinds of market information to help make investment decisions. These different types of information are generated in the intersection between various layers of activity—or rings. For example, the **securities firm** sits at the intersection between the buyer and seller of a particular stock, often playing an important role in controlling the supply and the demand for that stock. The **CEO** of a company sits at the intersection of the company's operations and its stockholders. The **customers** sit at the intersection of the company's services and its own bank account, which it can choose to spend on the company's products. Similarly, the **suppliers** sit at the intersection of its own products and services and the company's cash. The **employees** trade their time for cash and shares of the company. The **government** uses its power to bound the company's conduct in the product, employment, and securities markets, to set standards for performance reporting, to take a piece of its cash flows through taxes, and to set the rate at which it can obtain capital through the control of interest rates.

Each of these rings generates information that results from collisions at the points of intersection. This information can be useful for investors. In fact, the more potential information from these six points of intersection that an investor can analyze, the better the quality of the investor's decisions.

Investors should keep in mind that this information is of relevance to any company, including Internet companies. As we will see later in this book, the Internet can be a useful source of information about these questions—as they pertain to Internet and other companies. As Table 1-1 indicates, the six points of intersection generate an ongoing stream of useful information that differs by ring. These rings of information have great significance to Internet investors. As we will examine later, an important part of successful investing in Internet stocks is the ability to find effective ways to get useful information on these topics. Through a combination of talking to the right people and monitoring the right Web sites, it is possible to obtain a significant amount of this information.

TABLE 1-1. RINGS OF INFORMATION DRIVING INTERNET STOCK PERFORMANCE

Ring	Questions of Value to Investors
Securities firm	Who is buying and selling shares of the company's stock?
	How large are the transactions on the buy and sell side?
	What are the significant short-term trends in trading volume on the buy and sell sides?
	What are the earnings expectations for the firm, and how likely is it that these expectations will be exceeded?
CEO	What new products or business alliances are under way that are likely to enhance the company's earnings?
	What problems is the company encountering in meeting its earnings targets, and how likely is it that these problems will cause the company to miss earnings expectations?
	Is the company gaining or losing market share? Why?
	Are there new technologies emerging that are likely to derail the company's long-term revenue stream? If so, what is the company doing about the challenge?
Customers	Are the customers likely to increase or decrease their overall budget for products in the firm's category? Why?

Ring	Questions of Value to Investors
	Are the customers likely to spend more or less of their budget on the firm's product? Why?
	Are the customers becoming more or less satisfied with the firm's products/services? Why?
	Are the customers currently reviewing competing products/services? If these products/services are competitive, how is the firm taking action?
Suppliers	Are the suppliers meeting their commitments to the firm in delivering their product or services in the requested quantity, at the expected quality levels, and for the anticipated price?
	If not, why not, and what is the supplier doing to improve its performance?
	Is the firm paying its suppliers on time? If not, why not?
	Are suppliers investing to improve the quality of the products and services it offers the firm? If so, what are the tangible benefits of these investments?
	Is the firm currently reviewing products/services from other suppliers? If these products/services are competitive, how are the firm's current suppliers taking action?
Employees	How do the firm's employees compare to those of its competitors in terms of intellect, creativity, ability to make independent decisions, and effective teamwork?
	What is the rate at which top people are being recruited by the firm relative to competitors?
	How satisfied and productive are the firm's employees relative to competitors'?
	How does the firm's turnover rate compare to competitors'?
Government	Is the government investigating any violations of securities laws or any "business conduct" laws by the company or any of its stakeholders?
	Is the government changing regulations that could have a significant impact on the company's operations? If so, what is the likely magnitude and direction of that impact?

Other crucial pieces of information, particularly the supply and demand for shares, is tightly controlled. In fact, I believe that this information is so central to making an informed investment decision that the government should require it to be reported to the public, so that the brokerage firms with access to the information do not enjoy an unfair advantage over other investors.

STRUCTURAL EVOLUTION OF NETWORKED BUSINESS

While the first two of our seven key concepts pertain to just about any business, we will explore later how they apply to investing in Internet businesses in much greater detail. The third concept—the structural evolution of networked business—is of specific relevance to investing in Internet business. We will explore this topic in much greater depth in Chapter 2.

To understand the concept of Internet business, it is crucial to recognize what the Internet is. The Internet is a network of networks. It was designed by the Defense Advanced Research Projects Administration (DARPA) starting in 1969 to keep U.S. computer operations running in the event that one or several of the network nodes were hit by a nuclear warhead. By creating a network of networks, the idea was that even if several of the nodes were destroyed, the remaining nodes could take over. From 1969 until the mid-1990s, people thought of this ARPANET—later the Internet—as a useful tool for communication between government and educational institutions.

In the mid-1990s, with the introduction of the Netscape browser and the emergence of the World Wide Web, people began to conceive of ways to use the Internet as a communications medium for business. As we will explore throughout this book, nine distinct Internet business segments have emerged. These segments have varying levels of profit potential, requirements for competitive advantage, and investment opportunity.

The point to make here is that all of these distinct segments share a common source of revenue growth. This source of revenue growth depends on how we think about the Internet as a source of revenues. At the end of 2000, for example, there were 122 million people connected to the Internet in the U.S. In theory, the Internet enables each of these 122 million people to be a buyer of products and services and a supplier of products and services. The fact that each person could be a consumer and a supplier serves to increase the number of potential transactions very dramatically.

To help think about the value of the Internet as a source of new business, we developed the concept of Gross Internet Product (GIP). GIP applies the concept of GDP to the Internet: GIP is the sum of all the units times the prices of all business conducted on the Web.

The growth of GIP is a function of four key growth drivers. The single most important driver of GIP is the percentage of e-commerce revenue generated by two-way network participants—e.g., network participants who are both buyers and suppliers. If the percentage of two-way network transactions increases from 1% to 10% of the total, GIP increases 900%. The second most important driver of GIP is the number of two-way network participants. For example, a 10% increase in the number of two-way network participants leads to a 21% increase in GIP.

Two growth drivers vary linearly with GIP. As the number of people connected to the Internet grows, so does GIP. Based on a model developed by Peter S. Cohan & Associates, a 10% increase in the number of people connected to the Web causes GIP to increase by 10%. Similarly, a 10% increase in the average transaction size leads to an 10% increase in GIP. As we will explore in Chapter 2, the biggest leverage point to increase the value of the Internet as a source of new business is to increase the proportion of two-way transactions.

Peter S. Cohan & Associates' model is based on a number of key assumptions, which we will discuss in Chapter 2. The point of the model is that it helps investors to realize which growth drivers are likely to have the most important impact on the value of Internet business more generally. Therefore, it will help investors to assess which industry segments are likely to have the strongest influence on the most critical growth drivers. As a consequence, investors may wish to focus their attention more closely on these segments.

NINE INTERNET BUSINESS SEGMENTS

An Internet business is a company that derives some or all of its revenues by virtue of the Internet. As we noted earlier, there are nine Internet business segments, each of which is a distinct industry with unique competitors, customers, profit dynamics, and requirements for competitive success. For those who have not read my book *Net Profit,* the nine Internet business segments are:

1. Network infrastructure
2. Internet venture capital
3. Web consulting
4. Web security
5. Web portals
6. Electronic commerce
7. Web content
8. Internet service providers
9. Web commerce tools

Let's define each of these Internet business segments in turn.

1. Network Infrastructure

Network infrastructure is the hardware that directs traffic over the Internet. Network infrastructure consists mainly of devices called routers, switches, hubs, bridges, and Network Interface Cards. Vendors of this equipment include Cisco Systems, 3Com, Juniper Networks, Extreme Networks, Cabletron, Nortel Networks, Lucent Technologies, and many others.

2. Internet Venture Capital

Internet venture capital firms provide capital, recruit managers, and help grow Internet companies so they can go public and generate high investment returns. Internet venture capital firms include CMGI, Internet Capital Group, Softbank, Kleiner Perkins, Sequoia Capital, and many others.

3. Web Consulting

Web consulting firms help organizations use the Web to improve their competitive positions. Web consulting firms accomplish this by first working with client executives to understand their business objectives. Then the Web consultants design and implement Web-based systems that help the clients achieve their objectives. Web consultants include Sapient, Scient, Viant, DiamondCluster International, Razorfish, and many others.

4. Web Security

Web security firms provide software and services that help protect organizations' information networks from unauthorized intrusion and tampering. Web security firms accomplish this by providing services such as ethical hacking, in which an authorized individual attempts to break into a firm's information network and thereby to identify security weaknesses. Web security firms also sell a variety of software products designed to plug such weaknesses. Web security firms include Check Point Software Technologies, Network Associates, VeriSign, and many others.

5. Web Portals

Web portals give Internet visitors a place to begin their exploration of the Internet. Web portals do this by offering search engines, electronic mail, information services, chat rooms, and other services. Some firms in this segment are attempting to make the Web their primary mode of transaction. Others are adding Web channels to existing conventional modes of business. Web portals seek to attract enough visitors so that companies will view the Web portal as an attractive place to advertise. Web portals include Yahoo, Terra Lycos, and GoTo.com.

6. Electronic Commerce

Electronic commerce (e-commerce) means selling products and services using the Internet. E-commerce firms let people trade securities, buy books and CDs, purchase computer hardware and software, obtain air tickets, reserve hotel rooms, conduct online auctions, and purchase many other products and services. E-commerce firms include E-Trade Group, Amazon.com, eBay, VerticalNet, and FreeMarkets.

7. Web Content

Web content firms produce news and analysis of the Internet. Web content firms hire reporters and consultants who collect information about the Internet. Web content firms transmit the Internet-related information and analysis through a variety of media, including magazines, newspapers, TV, radio, trade shows, and the Internet itself. Most Web content firms also produce content about other technologies besides the Web. Web content firms include CNET, Forrester Research, Gartner Group, and many others.

8. Internet Service Providers

Internet service providers (ISPs) provide individuals and organizations with connections to the Internet. ISPs use a variety of media to connect their customers to the Internet, including telephone wires, cable TV, regular TV, and (eventually) Low Earth Orbiting (LEO) satellite networks. ISPs include EarthLink, Microsoft Network, America Online, Excite@Home, Metricom, and many others.

9. Web Commerce Tools

Web commerce tools help organizations to conduct business over the Web. These tools include advertising management services and software, Web browsers, multimedia broadcast tools, search engines, and online catalog software. Web commerce tool vendors include DoubleClick, NetGravity, Allaire, Net Perceptions, Macromedia, Inktomi, Open Market, and many others.

For investors, these nine Internet business segments have tremendous significance. The reason is that investors can improve their chances of picking a winning company simply by choosing the right segment or segments in which to invest. For example, Peter S. Cohan & Associates' index of Internet stocks in these nine segments rose 339% in 1999. This overall increase, while extraordinary compared to the record 86% increase in the NASDAQ, still masks wide variations in returns across the nine segments. For instance, the stocks in the Internet venture capital sector increased 1,005% in value during 1999, while the Web content firms rose a relatively meager 32%. Similarly, the 67% drop in this index between December 1999 and December 2000 masks wide variations among segments. For example, during this period Internet venture capital lost 97% of its value, while Web security lost 26%. Chapters 3 through 12 explore these differences in much greater depth.

FOUR TESTS OF A WINNING INTERNET COMPANY

A successful investor in Internet stocks needs a way to sort through the nine segments and the companies within the segments in order to assess the relative value of a company in relation to its peers. Simply put, an investor needs the tools to pick the winner in the category. With these tools in hand, an Internet business strategist can likewise have a better chance of building a valuable Internet business.

To select winning Internet businesses, investors must filter out a lot of noise. For every 100 business plans, they may pick one in which to invest. An affirmative answer to each of the following four questions can help filter out the other 99:

- **Industry:** Does the firm participate in an industry with economic bargaining power?
- **Strategy:** Does the firm offer its customers a "closed-loop" solution?
- **Management:** Does the firm's management have integrity and the ability to adapt effectively to change?
- **Brand:** Has the firm assembled a compelling brand family?

The odds of being a winner increase dramatically if a firm picks the right industry in which to compete. Here are four questions to use in testing whether an industry has economic bargaining power:

- Does it target a large market, preferably one with over $5 billion in revenues?
- Is it able to charge a price that well exceeds costs?
- Does it offer a product or service in limited supply?
- Is it able to generate demand from powerful decision-makers?

Having picked the right industry in which to compete, a firm can increase its odds of becoming the market leader if it executes the right approach to creating customer value. In many industries, the leading company assembles disparate products and services into a package that solves customers' problems (a "closed-loop" solution). This approach contrasts with open-loop-solution firms, which offer pieces of the solution and leave the customer to select and integrate the products or services.

Two management qualities are critical to creating a winning Internet business. First, management must have absolute integrity. Obvious things to look for are absence of a criminal record, a good credit history, and good references. In the Internet business, it's essential that managers meet an even higher standard of integrity: They must consistently deliver on their commitments. Winning depends on trust, and trust depends on meeting commitments.

Management must also adapt to change. A good example is Yahoo, which evolved from a cataloger of Web sites into a global media company, keeping its lead as competition and customer needs changed.

The final test comes in the firm's ability to assemble managers, investors, customers, alliance partners, and underwriters whose brands stand for leadership in their respective fields ("the brand family"). Such a brand family conveys to an investor—who may not have a detailed understanding of the firm's market, technology, or organization—that a collection of highly respected people and organizations has concluded that association with this startup will enhance their reputations (and net worth).

Check Point Software, to which we referred earlier, passes the four tests this way:

Industry: Check Point Software targets the multibillion-dollar market for Internet security software, a market that is willing to pay for a solution it perceives as effective to defend against security breaches.

Strategy: Check Point Software's product line is driven by its market leadership in firewalls and is further bolstered by its alliances with other Internet security software firms, thus assuring customers that Check Point's products will work with others products, generating a greater sense of security for clients.

Management: Check Point Software's management team has consistently delivered very aggressive sales and profit growth even as the competitive environment has changed.

Brand Family: Check Point Software counts worldwide industry leaders among its customers and has partnered with important participants in all the key Internet security product areas.

INTERNET INVESTMENT DASHBOARD

Having picked a firm that appears to be a winner in an attractive industry, a successful Internet investor needs a way to monitor that company's per-

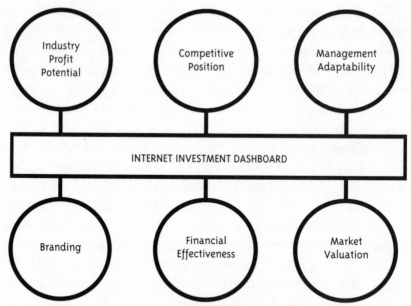

Figure 1-1. Internet Investment Dashboard

formance to determine whether or not to continue holding the stock. The Internet Investment Dashboard, as depicted in Figure 1-1, is particularly useful when an Internet company's stock price begins to drop.

Sometimes a drop in price may be attributable to "technical factors," such as insiders selling a stock with limited float during a time window mandated by the SEC. Simply put, senior executives have a limited period of time during the year when they can sell their stock. Often, such executives are selling the stock in the interest of not having all their "eggs in one basket." If a significant share, say 80%, of the stock in a publicly traded company is held by insiders, such selling can cause a significant drop in the stock price.

For short periods of time, it is common to observe significant gaps between the behavior of a stock and the performance of the business. If a stock price's drops are not related to a deterioration in the company's industry, competitive position, or management, then the drops may represent an excellent opportunity to invest. Conversely, if a stock is rising in price, yet its industry, competitive position, or its management are deteriorating, the price rise could represent an excellent selling opportunity.

Over a longer time period, it is reasonable for investors to expect that the stock price and the performance of the business will align themselves.

As we will explore in Chapters 3 through 12, the business-performance indicators that investors should monitor are different for each of the nine Internet industry segments.

While the specific measures vary by industry segment, following are six classes of business-performance indicators that investors should monitor for the stocks that they are holding. Ideally, investors should track these indicators for the company and for all the publicly traded companies in that Internet market segment:

- **Industry profit potential** indicators help investors determine whether the industry's overall returns are likely to increase or decline. Specific examples of such measures include revenue growth rates, profit margin trends, market share distribution, changes in the number of competitors, and the height of entry barriers.
- **Competitive position** indicators help investors determine whether a specific company's competitive position is improving or eroding. Specific examples of such measures include relative market share changes, changes in the level of customer satisfaction with the company, relative rates of employee turnover and hiring, rate of new product introduction, and changes in analysts' perceptions of the company.
- **Management adaptability** indicators help investors assess how well the firm's management is adapting to change in the industry. Specific examples of such measures include analysis of management's previous experience adapting to change, the rate at which the company makes acquisitions, effectiveness of post-acquisition integration, the rate at which the company forms strategic alliances, the company's method of monitoring customer feedback, new competitors, and new technologies.
- **Branding** indicators help investors assess how well the firm is branding itself in the minds of investors and customers. Such measures include analysis of the track records of the firm's venture capital investors, investment bankers, management team, partners, and customers.
- **Financial effectiveness** indicators help investors to assess how well the firm is managing its resources. To measure financial effectiveness, it is particularly useful for investors to track specific indicators over time relative to those of key competitors. Specific indicators worth watching are likely to vary by segment, but they might include marketing expenses per new customer, research and devel-

opment expenses as a percent of new-product revenues, gross profits per customer, revenues per employee, and cash-burn rate. Deterioration in such measures relative to competitors should indicate a potential problem with the business that may serve as an early warning indicator;

- **Market valuation** indicators may help investors assess whether a particular company's stock in an industry is under- or overvalued relative to its peers. When analyzed in the context of the other five indicators, the market valuation indicators can help assess important investment timing decisions. Specific market valuation indicators should be assessed in the context of competing firms over time and include market-capitalization-to-sales ratios, revenue-growth-to-market-capitalization-to-sales ratios, market-share-to-market-capitalization-to-sales ratios, and price/earnings ratios to earnings growth rate (for profitable Internet industry segments).

SEVEN-STEP INTERNET INVESTMENT PORTFOLIO-MANAGEMENT PROCESS

In order to make use of the foregoing concepts, investors need to follow a systematic process for managing their portfolio of Internet investments. The process we will develop in this book is designed to achieve the following:

- Match investment decisions with investor objectives and tolerance for risk;
- Respond effectively to sudden changes in market conditions; and
- Simplify personal financial management.

The process for managing a portfolio of Internet investments is similar in many respects to the process of managing any portfolio. The specific elements of the process we will develop in Chapter 13 are intended to help make the process of investing particularly effective for Internet stocks.

The process includes the following steps:

- **Identify investment objectives and risk tolerance.** The process begins by estimating specific amounts required to pay for a home, education, or retirement, combined with an analysis of the level of risk that the individual is willing to take on to achieve these objectives. The outcome of this process is a decision as to whether or not it makes sense for the individual to invest in Internet stocks and, if so, the appropriate level of investment in Internet stocks.

- **Determine the suitability for private- versus public-equity investing.** Based on the level of risk tolerance, required return, and financial resources, the investor must decide whether or not to invest in private or public Internet equities. For those who can invest in private equities, this step will determine the extent of the individual's portfolio that should be invested in private equities.

- **Select Internet industry segments that are most likely to meet investor objectives and risk tolerances.** For those who determine that Internet stocks fit their objectives, the next step is to identify the specific segments that are most likely to help meet the investor's objectives. This can be accomplished through a combination of analysis of the Internet sector's past performance and a forecast of how returns in that segment are likely to evolve. The outcome of this step is the identification of a specific set of Internet industry segments that most closely align with the investor's objectives and risk tolerances.

- **Pick the most attractive companies in the selected Internet industry segments.** This step applies many of the techniques we discussed earlier to enable the investor to assess which companies are most likely to emerge as winners in the selected segments. By analyzing the company's strategy, management, brand family, and financial management, the investor can assess the relative strength of the firm and thus its potential price appreciation. The outcome of this step is a list of companies in which to invest at the appropriate time.

- **Assess the appropriate timing of the purchase.** This step in the process is intended to help investors assess the appropriate timing for the investment in the selected companies. While we have a bias toward investing in leading companies in attractive industries, analysis shows that timing can help improve returns. For example, if a leading firm with strong fundamental business performance experiences a significant decline in price, this decline could be a good buying opportunity. The outcome of this step is a decision about the timing of the purchase of the stocks selected in the previous step.

- **Monitor industry, company, and market performance.** Once the purchase decision has been made, the investor must monitor developments in the industry, the company, and the overall securities markets in order to decide whether to purchase more shares of the securities in the portfolio, sell the shares, or otherwise rebalance the portfolio. In Chapter 13, we will define different kinds of information that need to be monitored and the frequency with which these items must be monitored in order to provide investors with the information they need to make informed decisions. The outcome of this step is the Internet Investment Dashboard (IID).
- **Adjust portfolio to achieve objectives.** Based on the information in the dashboard—and possible changes in investment objectives and risk tolerances—the investors can make the appropriate adjustments to their Internet investment portfolios.

CONCLUSION

Investing in Internet stocks can be done in a rational way to achieve investor objectives within identifiable risk tolerances. To understand the method, read on.

Chapter 2

The Value of the Network

IT HAS BECOME almost a matter of Internet legend that the spur to Amazon.com founder Jeff Bezos's ambitious drive from New York to Seattle was the 2,300% annual growth of the Internet back in the mid-1990s. Bezos was convinced that there was no other business opportunity on the planet that came even close to growing that fast. Even today, the valuations of Internet companies depend—in part—on firms' ability to maintain the fastest revenue growth rates on earth.

In order to put these growth rates into context, this chapter explores some theoretical questions that bear directly on how we can quantify the value of a network from a business perspective. In this exploration, we will define different network business configurations. We will estimate the business value of these different configurations. Further, we will examine the specific factors that drive changes in the value of such networks. We will conclude by articulating the principles that emerge from this evaluation to drive value creation. These principles will prove useful as we determine how best to profit from investing in companies whose growth depends on increased network value.

NETWORK TYPES

The basic concept of a network is fairly straightforward—a collection of nodes that connect to each other in different ways. From a business perspective, the nodes are often thought of as sellers and consumers. An example might be a Web site such as Amazon.com—a seller node—connected to 17 million consumer nodes. While this configuration represents a simplification of the reality, it does illustrate the concept of a one-to-many network. As Figure 2-1 illustrates, in a one-to-many network, one seller offers products or services to many buyers. This one-to-many network has five one-way transaction paths. In general, a one-to-many network with one seller and *n* consumers will have *n* one-way transaction paths.

Such one-to-many networks contrast with many-to-many networks (see Figure 2-2), in which all network participants can be both buyers and sellers. Online auctions such as eBay provide a useful example of a many-to-many network. Here each auction participant can be a bidder during one auction and a seller in another auction. In fact, each many-to-many network could be the site of several auctions in which a participant is a buyer at the same time that the participant is selling several items.

In some sense, this concept of the many-to-many network is as old as populism and the notion of the small business. The key difference here is that the Internet has the ability to increase the number of network participants. Beyond increasing the number of many-to-many network participants, the Internet unlocks a tremendous amount of individual creativity by simultaneously lowering the cost of jumping into the network and widening the potential market of buyers.

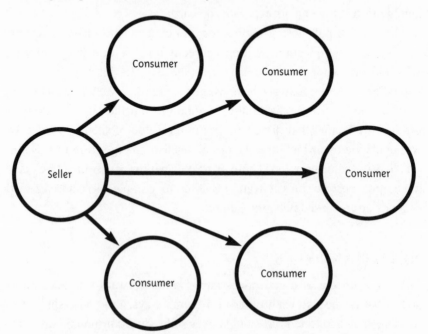

Figure 2-1. One-to-Many Network with One Seller and Five Consumers

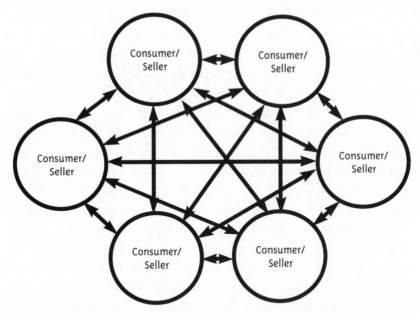

Figure 2-2. Many-to-Many Network with Six Consumers/Sellers

The mathematics of many-to-many networks helps explain their enormous potential for creating economic growth. In the example depicted in Figure 2-2, which has six consumers/sellers, there are 15 two-way transaction paths. The general formula for the number of two-way nodes in a many-to-many network of n nodes is $[(n-1) \times n]/2$. If we assume that a one-way transaction path can handle half as many transactions as a two-way transaction path—and hold all other factors equal—the potential value of the many-to-many network in Figure 2-2 is six times that of the one-to-many network in Figure 2-1.

In general, again holding all other factors equal, a many-to-many network with n nodes is worth n times minus one that of a one-to-many network with one seller and n consumers. This is a particularly stunning result if we consider the implications for the Amazon.com network with its 17 million consumers. If all those consumers and Amazon.com sold and bought from each other, the value of that network would be 16,999,999 times greater than Amazon.com as it is currently operated. While Amazon.com's 17 million consumers constitute 17 million one-way transaction paths, if Amazon.com were a many-to-many network, it would have 1.444999×10^{14}—or about 14 quadrillion—two-way transaction paths.

Simply put, if Amazon.com could get its 17 million visitors to buy and sell from each other—and then take a piece of all that new business—Amazon.com would earn a lot more revenue than it does today.

While these theoretical concepts simplify reality, they also provide an interesting framework to use in thinking about where the Internet came from as a source of business value and where it is heading in the future. The Internet is currently much closer to a one-to-many network than to a many-to-many network. Nevertheless, the concept of online auctions popularized by eBay is much closer to the many-to-many model. And even one-to-many model firms such as Amazon.com create a bit of interaction among their consumers through the concept of purchase circles, in which people with common interests share their insights with each other (in general, a cashless transaction).

Another important point to be drawn from this example is that these kinds of networks existed, at least in theory, before the Internet. Theoretically, any person could buy and sell from any other person. With the advent of the telephone, the cost and time required to conduct such transactions dropped significantly from the era when people communicated via letters and telegraph. The Internet simply represents a quantum improvement in the cost and time to achieve the level of information-sharing that would be required to realize the theoretical business value inherent in a many-to-many network. By lowering the cost and time to share such information, the Internet unlocks a significant amount of theoretical business value. As a consequence, our economy experienced an unusually high rate of growth in the late 1990s.

There remain significant barriers, however, to unlocking the full potential of a many-to-many network, and these barriers are not likely to fall in the immediate future. Such barriers include the fact that the cost of devices to connect people to the Internet remain too high to make them useful for all people. In addition, different people have different levels of producing and consuming capacity, and thus there are likely to be limits to the size of such many-to-many networks. Nevertheless, as the cost of connecting people to the Internet declines, it is likely that significant unrealized economic value will be unlocked.

This discussion sets the stage for an analysis of the value of network business. What we will find is that many of the Internet stocks that some analysts typically describe as grossly and inexplicably overvalued are in fact significantly undervalued in light of the business value that the Internet has the potential to create. As a consequence, this theoretical discussion has significant importance insofar as it provides the conceptual underpinning for the analysis of the values of Internet stocks, which we will examine in subsequent chapters.

NETWORK VALUE

We will address the value of the network as a source of new business from the perspective of where it was in 1999 and where it could be in the 21st century, based on a specific set of assumptions. According to a University of Texas study published in 1999, the Internet contributed $300 billion to the U.S. economy (University of Texas at Austin, 1999). This total includes the nine Internet business segments we introduced in Chapter 1. It is useful to think of the segments in two categories: e-commerce and everything else.

The "everything else" can be thought of as **complementary goods**. An example is that of a home purchase. Every time a new home is purchased, there is likely to be a set of related purchases of items such as furniture, carpeting, etc. These related purchases are complementary goods. As Figure 2-3 illustrates, it is interesting to note that for every dollar of e-commerce, there is $1.49 worth of purchases of Internet infrastructure. The figure similarly illustrates the proportions of the seven other Internet business segments that are attributable to a dollar's worth of e-commerce.

Another way to think about the relationship between e-commerce and the other Internet business segments is to imagine a V of migrating geese. E-commerce is the lead goose and the progress of the other segments, such as Web consulting or Web advertising, depends on the rate of travel of the lead goose. When dot-coms fell out of the sky in April 2000, the other segments followed within a few months. The recovery of the Internet economy will depend on how effectively land-based businesses adopt e-commerce into their operations.

Figure 2-3 suggests that there is a rough correlation between the relative size of an Internet business segment and its relative profitability. Internet infrastructure has the highest inherent profitability of the Internet business segments, while Web content has the lowest profit potential. Here

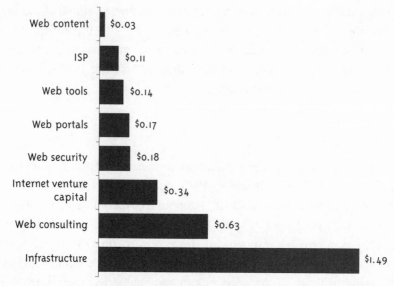

Figure 2-3. Internet Economy Segment Revenue per Dollar of E-Commerce, 1999
Source: Peter S. Cohan & Associates

the willingness to pay lots of money seems to correlate with the value to buyers.

While the Internet economy reached significant scale on an absolute basis in 1999, it still represented a relatively small proportion of overall GDP. For example, 1999 U.S. GDP totaled $8.6 trillion, so the $300 billion Internet economy represented a mere 4% of that amount. Therefore, it is safe to conclude that the amount of hype associated with the Internet economy exceeds its size in proportion to the economy.

The future value of networked business, however, according to the assumptions used in our model, is enormous. The critical assumptions used in this model include the following:

- 273 million U.S. citizens have Internet access. This represents 100% of the current U.S. population, although it is anticipated to represent less than the total U.S. population in the future. Since 122 million U.S. citizens currently have Internet access, the "virtually complete" access assumption can only be valid if people begin to access the Internet via TV, cable, telephone, and/or PC. Internet industry leaders are working toward the realization of this assumption. However, there are many barriers, which we will discuss further.

- The average amount of e-commerce conducted per one-to-many network node remains the same in the future as it was in 1999, $667 per node per year.
- The average amount of e-commerce that will be conducted per many-to-many node will be $30 in the future.
- An estimated 99% of e-commerce will continue to be conducted through one-to-many networks, while 1% of e-commerce will be conducted by many-to-many networks.
- The proportions of the complementary goods will remain the same in the future as they were in 1999.

Later, we will test these assumptions by estimating the percentage change on GIP of a 10% change in each of the critical assumptions. Nevertheless, these assumptions are more likely to underestimate than overestimate the future levels of GIP.

Using these assumptions, the future level of U.S. GIP is likely to be an enormous number—over $7.5 quadrillion. The 1999 U.S. GDP represents a bit more than 0.1% of this amount. The $1.5 trillion market capitalization of Internet stocks noted earlier represents an even smaller fraction, 0.02%, of this $7.5 quadrillion in future GIP. If this estimate of future GIP is even remotely close to being correct, it becomes more plausible to anticipate higher stock market values for firms that exploit this enormous GIP potential in a profitable fashion.

The reason for this becomes clearer as we look at how the nine components of GIP would be distributed. Figure 2-4 presents the results of how the $7.5 quadrillion will be distributed.

Another way to look at these numbers is through the compound annual growth rate required to reach the $7.5 quadrillion worth of GIP in various future periods. If we assume, for example, that it takes 15 years to reach this amount, then GIP would need to grow at a 96% annual rate between 1999 and 2014. Figure 2-5 details the range of growth rates that would be required under varying assumptions about how long it would take to reach $7.5 quadrillion in GIP. Since the Internet has been doubling every year for the last several years, its growth rate would need to be sustained for the next 15 years in order for GIP to reach $7.5 quadrillion by 2014.

The law of large numbers suggests that such rapid growth rates will not be sustained. It is possible that the full economic potential of Internet business will be realized in stages as the incremental costs of going online

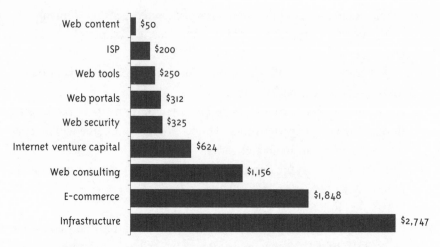

Figure 2-4. Internet Segment Revenues at $7.5 Quadrillion of GIP, in Trillions of Dollars
Source: Peter S. Cohan & Associates

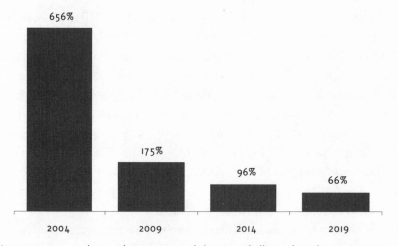

Figure 2-5. Annual Growth Rate to Reach $7.5 Quadrillion of GIP by Various Years
Source: Peter S. Cohan & Associates

drop—and the offsetting benefits increase—for the more Internet-resistant market segments. Despite the slowing growth that is likely to occur, the Internet remains one of the fastest-growing technologies in history in terms of its adoption, as Figure 2-6 illustrates.

As the chart suggests, the personal computer is the only new technology that has been adopted anywhere near as quickly as the Web. The Web's rapid adoption rate might prepare us intellectually for its enormous future impact on our economy. Emotionally, however, it still creates discomfort

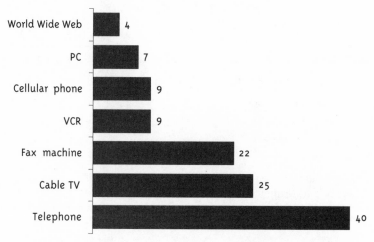

Figure 2-6. Years to Reach 100 Million Users, by Technology
Source: Booz, Allen & Hamilton

to contemplate its potential magnitude. To help understand the factors driving the impact of the Internet on the economy, let's evaluate the factors that drive its magnitude.

NETWORK VALUE DRIVERS

A fairly limited set of variables has a tremendous impact on GIP. As we will see, this analysis can provide investors with useful insights for investing. Whichever firm is doing the most to influence the most powerful drivers of increased GIP is likely to capture the most value for its shareholders.

As Table 2-1 indicates, the single most significant driver of GIP is the number of many-to-many transactions on the network, driven by the number of individuals connected to the network, or the number of nodes. As the table indicates, a 10% increase in the number of people connected to the network increases GIP by 21%. This powerful economic leverage is the driving force behind the efforts under way within the Internet industry to increase the proportion of the population that is connected to the Internet. All the investments being made to connect more people through paths other than PCs are intended to increase the proportion of people connected to the Web, because increasing this number is by far the most significant driver of increasing the value of Internet business.

TABLE 2-1. DRIVERS OF GIP

Variable	Initial Value	10% Increase	% Chg. in GIP
Number of nodes	111 million	122 million	21%
Number of many-to-many paths	6,160 trillion	6,777 trillion	10%
Percent of e-commerce from many-to-many paths	1.0%	1.1%	10%
Average annual transaction amount per many-to-many path	$30	$33	10%
Average annual transaction amount per one-to-many path	$667	$733	0%

Source: Peter S. Cohan & Associates

The proportion of e-commerce from many-to-many nodes is also an important driver of GIP. The relative importance of this factor is masked somewhat by its currently small proportion of the total. So a 10% increase in the proportion of many-to-many transactions from 1.0% to 1.1% increases GIP by 10%. Similarly, an increase of 10% in the number of many-to-many nodes, holding the proportion of the total constant, leads to a 10% increase in GIP. These findings have an important implication for investors—namely, that the online auction concept has enormous potential to unlock value. Therefore, investors should focus their attention on firms that are likely to be leaders in the online auction market and thus can drive an ever-higher number of many-to-many transactions.

GIP is also highly sensitive to an increase in the average annual dollar amount of many-to-many transactions. As Table 2-1 indicates, a 10% increase in this value, from $30 to $33, causes GIP to increase by 10%. If the $30 assumption turns out to be much too low, the value of GIP could increase linearly with the increase in the average dollar amount of the many-to-many transactions.

Finally, it is interesting to note that an increase in the average annual amount of one-to-many transactions has virtually no impact on GIP. The table notes that a 10% increase in the average annual one-to-many transaction, from $677 to $733, results in no increase in GIP. Actually, the impact is about 0.01%, but close enough to zero to highlight the relatively trivial impact of increasing the size of one-to-many transactions.

Simply put, beyond increasing the absolute number of nodes on the network, the single most important factor driving increases in the value of Internet business is to maximize the number of many-to-many e-commerce transaction paths. Firms such as eBay, VerticalNet, Free-Markets, and Ariba are just beginning to tap into the enormous potential value of many-to-many e-commerce. The firm that is able to increase the level of many-to-many e-commerce and to capture the largest share of this unlocked value will earn enormous shareholder returns.

PRINCIPLES OF NETWORK VALUE CREATION

This model of network value creation can help investors screen for investment opportunities. Investors who look for companies that follow five principles of network value creation are likely to be rewarded with superior investment returns. As we will see, merely finding companies that follow these principles is not sufficient. However, finding such companies represents an important first step toward picking winning Internet investment opportunities. In particular, investors should seek out companies that follow these five principles:

1. **Increase the number of people connected to the network.** Investors should look for companies that are taking actions to increase the number of people connected to the Internet. Firms seeking ways to connect people to the 'Net through non-PC devices—Microsoft, Palm Computing, Nokia, and others—are among the candidates that recognize the immense value creation that is likely to result from increasing the number of people connected to the Web.

2. **Encourage an increase in the number of many-to-many Internet transaction paths.** While the proportion of e-commerce that takes place through many-to-many transaction paths is currently about 1%, the companies that are seeking to increase the number of many-to-many paths are most likely to take a significant share of the increase in value that they are helping to create. Firms such as eBay, FreeMarkets, Ariba, and Commerce One are among the increasing number of firms that are seeking to increase the amount of auction business on the Web. Simply put, by realizing the potential that each individual can be both a buyer and a seller, winning Internet companies will unlock tremendous economic potential.

3. **Capture the complementary goods.** Firms that recognize the need for $1.49 worth of infrastructure for every dollar of e-commerce are among the ones that are likely to grab a huge share of GIP. Clearly, firms such as Cisco Systems, Juniper Networks, EMC, Akamai Technologies, and Oracle have recognized the enormous market opportunities induced by the ongoing growth of e-commerce. Investors who ignore the value of infrastructure are likely to forgo significant profit opportunities.

4. **Identify and overcome barriers to growth in the number of people connected to the Internet.** There are enormous reservoirs of fear and resistance to change within the economy. Internet businesses that take an active role in setting an example of how to overcome that resistance to change are likely to benefit enormously. For example, firms such as Charles Schwab have set an example of how to reinvent themselves in order to profit from the transition to e-commerce. Investors should seek out firms that are able to overcome these barriers because their leadership suggests tremendous shareholder return potential.

5. **Prevent external forces from impeding the growth of Internet business.** Beyond the fear of change, many external forces threaten to impede the growth of Internet business. These forces include the threat of taxation on e-commerce, security breaches from external and internal sources, the inability of the current Internet architecture to adapt robustly to the growth in demand for the Internet, and the inability of Internet service providers to provide sufficiently speedy and cheap connections to the Internet. Firms that help to reduce these growth impediments are likely to benefit from the growth they unlock.

CONCLUSION

Despite the enormous shareholder wealth that has been created and destroyed in the last six years owing to Internet business, we have yet to tap even 1% of the full potential of the Internet. If the value of Internet business continues to grow at its historical rate—roughly, doubling every eight to 12 months—it is possible that the potential value of Internet business could be realized within 15 years. Investors can tap into this value by identifying companies that are pushing the hardest to increase the number of people connected to the Internet and to encourage the growth of many-to-many transaction pathways. As 2000's 67% drop in Internet stocks suggests, it is crucial to make careful investment choices among these companies. In the next chapter, we will explore the growing divide between the companies that are participating in this growth and the vast majority of companies, which are taking a more measured approach to capturing the Internet's business value.

Chapter 3

INTERNET STOCKS AND THE GENERAL MARKET

B EFORE WE EXPLORE how investors can profit from Internet infrastructure, we need to step back and examine the concept of the nine Internet business segments that were first presented in my 1999 book *Net
Profit*.

NINE INTERNET BUSINESS SEGMENTS

While the media has devoted the most attention to firms such as Ama-
zon.com, successful investors in Internet business have realized that there
is much more to the Internet than e-commerce. As we will see throughout
this book, one of the most important choices an investor can make is picking the right Internet business segments in which to invest. While my book
Net Profit defined nine distinct Internet business segments, for the purposes of explanation it is useful to think of Internet business in terms of
the 1850s gold rush.

In the gold rush, there were people who sold gold miners the picks and
shovels, there were the providers of transportation from mines to markets,
and there were the gold miners themselves. Gold rush lore has it that the
most profitable companies during the period were the providers of picks and
shovels. As we will see, the analogy between the gold rush and the Internet
holds as to which businesses have ended up earning most of the profits.

The nine Internet business segments can be thought of as fitting into
one of three broad categories.

Picks and Shovels

Who is providing the picks and shovels of the Internet age? Six of the nine Internet business segments fit within this category. The Internet business segments in the picks and shovels category include Internet infrastructure, Web consulting, Internet venture capital, Web tools, Web security, and Web content. These six businesses make up an essential part of the list of supplies required by an e-commerce firm.

To begin with, an e-commerce firm generally needs to raise venture capital in order to implement its business plan. With capital in hand, the e-commerce firm may hire a Web consultant to help design and program its Web site. In order to build and operate the site, the e-commerce firm needs to purchase Web tools, Web security software and services, and Internet infrastructure. Once the Web site is up and running, the e-commerce firm needs to make potential customers and partners aware of the site with the help of Web content providers.

Transportation

During the gold rush, the miners needed a way to get their gold to the market in order to cash in on their gains. While the analogy is a bit rough, one Internet business segment, the Internet service provider, can connect businesses with their customers. The analogy is rough because some e-commerce providers choose to host their own Web site, in which case they control their own "transportation" mechanism. Other firms choose to outsource the operation of their Web sites, either by hiring a business Internet service provider, such as Exodus Communications, or by running their site on an Internet service provider such as EarthLink or TheGlobe.com.

Gold Miners

While only two of the nine Internet business segments fall into the "gold miner" category—e-commerce companies and the Web portals—these two segments have received the lion's share of media attention since Netscape's initial public offering in 1995. Over the last several years, E-commerce and Web portals have evolved substantially from their initial focus on consumers.

The initial goal of e-commerce firms was to provide many free services in order to attract a significant audience. By attracting a significant audience, the e-commerce firms hoped to generate advertising revenues that would offset the high advertising and sales costs required to attract the large audiences. In late 1999, it became clear that such advertising-based strategies would not work and that even the combination of advertising and e-commerce revenues would not be sufficient to generate positive cash flow.

Similarly, Web portals initially gave away a significant number of free services, including Web searching, news, stock prices, electronic mail, and instant messsaging. The Web portals hoped to attract large audiences and monetize the traffic by selling advertising. The advertising-only strategy did not work for all participants and helped contribute to the consolidation in the industry. Recently, surviving Web portals have added e-commerce as a source of revenue, resulting in profitability for leaders such as Yahoo.

In the latter half of 1999, investors began to believe that the business-to-business (B2B) segment of e-commerce had greater profit potential. Claiming that businesses using the Internet to transact business with each other would generate $1.3 trillion in revenues by 2002, analysts generated significant investor interest in firms targeting this market. Companies like Internet Capital Group and VerticalNet attracted significant investor capital and rose rapidly in price. By the first quarter of 2000, however, large Old Economy companies made announcements that they would enter the B2B fray. The threat of these new entrants took away a significant amount of investor interest in the B2B "pure plays." The ultimate disposition of the profits in the B2B area is a subject for further analysis in Chapter 7.

The next nine chapters of this book examine the factors that drive the relative stock price performance of each of these nine Internet business segments.

Between 1998 and 2000, Internet stocks went up and down at more extreme rates than the major indices. For example, in 1999 the Dow Jones Industrial Average (DJIA) rose 20%, the NASDAQ was up 86%, and Peter S. Cohan & Associates' Internet Stock Index rose 339%. In 2000, however, the Internet Stock Index had lost 67% of its value, while the DJIA was down 6% and the NASDAQ had declined 39%. Over the two years, Internet stocks rose 6%, the DJIA was up 18%, and the NASDAQ had gained 13%.

To put that 300% return in perspective, it is important to understand that not all Internet stocks are created equal. In 1999, some sectors performed a lot better than others. Even within some sectors, there were wide gaps between the winners and the losers. The 339% gain masks very wide variations among the nine Internet business sectors that Peter S. Cohan & Associates tracks. For example, 1999's top performing sector, Internet venture capital, was up 1,116% for 1999, while the worst performing sector, Web content, grew only 22%. The index's 2000 performance similarly varied by sector.

Following is a review of the nine sectors.

Internet Venture Capital

Internet venture capital—investing in Internet startups—was up 1,116%. With only two stocks, Internet Capital Group (up 1,291%) and CMGI (up 940%), this index declined 97% in 2000. After April 2000, investors, abandoned publicly traded Internet venture capital firms because their portfolio companies, primarily in e-commerce, declined in market value. The outlook for this sector is negative for the next several years as venture capital firms take their losses.

Web Portals

Web portals—sites that serve as an entry point to the Web—were up 167% in 1999. The best performer was Yahoo (up 265%). Consolidation (as in 1999's merger between Excite and At Home) continued in 2000 with the merger of Spain's Terra Networks with Lycos and the merger of Go2Net with Infospace. The 2000 stock market performance of this sector was down 85% due to investors' fears of a dropoff in Web advertising.

Web Tools

The Web tools sector—software for building and operating Web sites—was up 448% in 1999. This sector included such strong names as Broad-Vision (up 1,494%) and DoubleClick (up 1,037%). Not all Web tools names did well, however. E-currency company CyberCash's stock dropped 35% as the market failed to adopt its service in sufficient numbers. This sector lost 81% of its value in 2000, particularly hurt by the drop in Web advertising stocks such as Engage, Net Perceptions, and DoubleClick.

Web Consulting

Web consulting—helping executives develop and build Web sites—was up 322% in 1999. The best performers were the consulting firms that focus exclusively on e-business services—Diamond Technology Partners (up 574%) and Proxicom (up 565%). Firms transitioning to e-business performed less well. For example, Cambridge Technology Partners was up 19%. In 2000, Web consultants lost 89% of their value as dot-coms and their land-based peers stopped clamoring for Web consulting services.

Internet Service Providers

The Internet service provider sector was up 366% in 1999. Paul Allen invested in Metricom, a wireless ISP, whose stock rose 1,416%. The long-term winners in this sector have been the business ISPs, such as Exodus Communications (up 1,006%). The losers are likely to be the consumer-oriented ISPs, such as EarthLink (down 25%). In 2000, ISPs lost 69% of their value due to the collapse of free ISPs such as Juno Online (down 98%) and troubled business ISPs such as PSINet (down 98%).

Web Security

In 1999, Web security—software and services that secure the Web—was up 263%. There were wide variations in performance between winners and losers in the segment. For example, VeriSign (up 1,192%) and Check Point Software (up 334%) were big winners as a result of their dominance in attractive market segments. Network Associates (down 60%) and Secure Computing (down 34%) both suffered from less than stellar management. In 2000, Web security lost 26% due to declines in Network Associates (down 84%) and VeriSign (down 61%), partially offset by gains in Check Point Software (up 169%) and JSS Group (up 10%).

Electronic Commerce

The big theme in 1999 electronic commerce was the split between business-to-business (B2B) and business-to-consumer (B2C). Electronic commerce overall was up 201% in 1999, masking the big differences between B2B and B2C. For example, B2B leader Commerce One was up 866% in 1999, while B2C poster child Amazon rose only 42%. Online CD vendor CDNow lost 53% of its value in 1999. By 2000, CDNow had run out of money and was acquired by Bertelsmann, the German media company. In 2000, e-commerce lost 76% of its market value thanks largely to the collapse of B2C firms such as money-loser Priceline, which lost 97% of its stock market value.

Infrastructure

Infrastructure—the hardware that directs network traffic—was up 144% in 1999. The best performers were firms like Redback Networks (up 289%) and Juniper (up 244%), whose products speed up connections to the Internet. Second-tier competitors like 3Com (up 5%) suffered in competition from Cisco (up 131%), while Cabletron (up 210%) benefited from new management. In 2000, infrastructure lost 36% of its value as firms like Lucent (down 81%) were slightly offset by gigabit router leaders such as Juniper (up 122%).

Web Content

In 1999, the Internet sector laggard was Web content—firms that report on Web activities—which grew a paltry 22%. In fact, if we take out two winners, CNET (up 326%) and Forrester Research (up 58%), all the other firms were flat or lost market altitude in 1999. For example, research firm Meta Group saw its stock price drop 36% in 1999. The Web content sector lost 47% of its value in 2000, and prospects for this section remain uninspiring.

This sector-by-sector analysis raises some deeper questions. For example, how do Internet stocks perform relative to the DJIA and NAS-DAQ indices in periods of increasing and declining stock prices? What are the factors—such as interest rates, earnings growth rates, money flows—that drive the differences in performance between Internet stocks, the DJIA, and the NASDAQ during these up and down periods? What do leading analysts and industry executives think drives the relative performance of Internet stocks? What general principles emerge from this analysis that can help the average investor?

INTERNET STOCK RELATIVE PERFORMANCE IN UP- AND DOWN-MARKET PERIODS

Compared to the other stock indices, Internet stocks went up more in up markets and down more in down markets. In 1999, compared to the NASDAQ and the DJIA, Internet stocks surprisingly tended to drop less in down markets and advance more in up markets. In March 2000, a shakeout began in which investors in Internet stocks decided to head for the exits—causing Internet stocks to lose more value than the NASDAQ and the DJIA.

We discern down-market periods by looking at recent history, defining down periods as the beginning and end of times when the DJIA lost value. Conversely, up markets were periods between the down periods—when the DJIA gained in value. As Figure 3-1 illustrates, between April 1999 and December 2000, there were a total of nine switches in direction for the DJIA. Four of these periods were down periods; five were up periods. The biggest down period for the DJIA was a 15% decline in value from mid-January to mid-March, 2000. During this down period, the NASDAQ and the American Stock Exchange's Internet index, known as the IIX, moved in the opposite direction from the DJIA, with the NASDAQ increasing 24%.

The DJIA outperformed the NASDAQ and the IIX in four time periods. In April/May of 1999, the DJIA was up 10% while the NASDAQ and IIX were up a mere 2% and 3%, respectively. In March/April 2000, the DJIA jumped 16%, but the NASDAQ and the IIX were down 14% and 6%, respectively. Between April and October 2000, the DJIA dropped 11% while the NASDAQ and IIX were down 24% and 25% respectively. Finally, between October and December 2000 the DJIA rose 8% while the NASDAQ and the IIX lost 22% and 35% respectively. This divergence between the Old and New Economy stocks suggests that investors tend to put their money into the New Economy stocks when they are afraid of losing out on growth opportunities, and they scramble into Old Economy stocks when they fear a drop and they want to preserve the gains they made in their New Economy stocks.

Interestingly, the IIX performed best during an up-market period, between October 1999 and January 2000, rising 67% while the NASDAQ climbed 45% and the DJIA rose 16%. This huge rise in the IIX was followed by a period from January to March 2000 during which the DJIA lost 15% of its value, the NASDAQ grew 24%, and the IIX grew 7%. The rea-

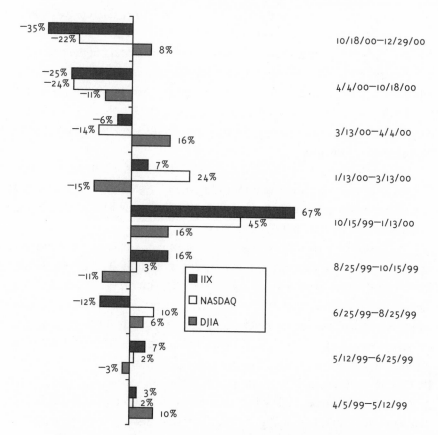

Figure 3-1. Performance of Dow Jones Industrials, NASDAQ, and
Internet Index (IIX) in Up- and Down-Market Periods, April 1999–December 2000
Source: PC Quote, Peter S. Cohan & Associates Analysis

son for the IIX's relatively weak performance is that investors were very
enthusiastic about biotechnology companies during this period because
they perceived that they would find a way to profit from the human
genome project. These biotechnology stocks made up a significant share
of the NASDAQ's relatively strong performance during this period.

The enthusiasm for these biotechnology stocks was muted by a joint
announcement from President Clinton and Prime Minister Blair of En-
gland, stating that the insights from the human genome project would be
made available to the public. By making this information public, the perceived profit that investors imagined would be flowing into the biotechnology companies quickly evaporated—thereby causing their stock prices
to fall suddenly.

The behavior of these indices relative to each other reflects two manifestations of investor fear. One manifestation is the fear of losing out on rapid growth, which leads investors to sell everything and put the proceeds into the most rapidly growing sectors. Another manifestation is the panic selling that accompanies drops in these rapidly growing sectors—a fear that leads investors to put their money into the DJIA, where they believe they have a greater chance of preserving the gains they have made in the fastest-growing sectors.

The behavior of these market indices highlights the potential decline of the effectiveness of the value investing belief system that we discussed in Chapter 1. Simply put, value investing applies to Old Economy stocks in many more cases than it does to New Economy stocks.

This split between Old and New Economy approaches was the catalyst for the March 2000 announcement by legendary fund manager Julian Robertson that he was closing down his hedge fund, Tiger Management. According to Reuters, Robertson, 67, told partners in a March 30, 2000, letter that he had liquidated Tiger's portfolio and was ready to return 75% of some $6 billion in investments to stakeholders in cash. Robertson decided to close the funds after losses suffered by investments in Old Economy stocks, which had dropped in value as technology stocks had grown more valuable. In his letter, Robertson noted that the demise of value investing and increasing investor withdrawals had created stress for Tiger investors and employees. Robertson also expressed his view that there was no real indication of an end to the shift away from his value investing approach (Reuters, 2000).

The stress that Robertson faced was measurable in a short, sharp drop in the value of the assets that his firm managed. For example, Tiger's assets dropped from $22 billion to $6.5 billion in 18 months. One reason for this huge decline was Tiger's 22% stake in U.S. Airways, whose stock lost 44% of its value between March 1999 and March 2000. U.S. Airways was a classic value "play" in that Robertson purchased a huge stake in the company after its stock had declined so far that Robertson believed the stock market was grossly understating the value that could be generated by U.S. Airways assets. Regrettably for Robertson, investors seemed more interested in putting their cash to work in rapidly growing technology stocks than in Robertson's assessment that U.S. Airways stock was cheap and therefore warranted accumulation to rectify the market's misvaluation.

Robertson's investment acumen seemed to misfire most profoundly in the last 18 months of his management tenure. In fact, Tiger's closure ended 20 years of 31.7% average annual returns, a record matched by few in the money-management world. Robertson had founded Tiger in 1980 with $8 million in assets. Tiger grew as Robertson gained a reputation as an astute stock picker. In 1996, Jaguar, one of Tiger's investment funds, posted returns of 50%, followed by a 72% return in 1997. Tiger's performance crumbled after an expensive and massive bet against Japanese stocks and the Japanese yen. Then, as Robertson stuck by Old Economy stocks, his performance lagged behind technology stocks (Reuters, 2000).

Robertson's experience points out that belief systems transcend even the most compelling of personalities. Simply put, if Robertson had been able to abandon his value investing approach and developed a method that made him comfortable investing in technology stocks, he would not have liquidated his portfolio. The challenging problem he faced was an unwillingness to abandon a belief system that had worked well for 18½ years. Robertson believed that the superior investment performance of technology stocks was a temporary aberration—not a fundamental shift that caused his beliefs to be out of step with investment reality and not vice versa.

Despite Robertson's experience, it is possible that the future will prove him right. Recent experience highlights enormous differences in market valuation between the New and the Old Economy stocks. The 2000 collapse suggests that these valuation differences reflected massive mispricing, not rational calculation. In early 2000, for example, *Barron's* argued that the U.S. had two distinct stock markets: the technology sector, in which many companies traded for over 100 times 2000 projected profits, and the rest of the market, where many stocks sold for as little as five times their annual earnings (Bary, 2000).

Investment veterans suggest that this two-tiered market was an aberration. For example, Jim Engle, chief investment officer at DLJ Asset Management, has noted that the Nifty-Fifty market of the early 1970s produced a handful of small companies trading for 100 times earnings. Now, Engle suggests, there are many big technology stocks trading at these valuations. Mark Boyar, head of a New York money-management firm, has pointed out that if a stock is not technology, Internet, or biotechnology, nobody wants it—a condition that Boyar notes he has never seen since he began managing money in 1969.

One indicator of the gap between the Old Economy and the New Economy has been the different performances of the DJIA and the NAS-

DAQ in the first two months of 2000. During those months, the DJIA fell 11% in part because it was pulled down by Old Economy stocks like International Paper, DuPont, United Technologies, Coca-Cola, and McDonald's. The NASDAQ, by contrast, rose 8%, with the help of Internet infrastructure stocks such as Intel, Cisco Systems, and Sun Microsystems.

The performance gap between the DJIA and NASDAQ was more firmly pronounced during the 12 months from February 1999 to February 2000, during which DJIA gained a mere 14% while NASDAQ increased in value by 102%.

But even the Dow includes technology companies such as Microsoft and Hewlett-Packard. So its overall performance does not fully reflect the declines experienced in a large number of Old Economy stocks, including the shares of regional banks, airlines, railroads, insurers, homebuilders, retailers, supermarkets, apparel makers, food companies, and manufacturers. Many stocks in these industries experienced price declines of 50% or more from February 1999 to February 2000. Technology, meanwhile, performed very well, driving the NASDAQ to rallies even as the Dow declined. The stocks of leading companies such as Cisco, Sun, and Oracle traded for 80 to 110 times their projected profits for 2000, while stocks in the Standard & Poor's 500 Index sold for 50 times 2000 anticipated earnings (Bary, 2000).

The April 2000 crash of Internet stocks precipitated a closer alignment of Old and New Economy stocks. In particular, some well-known Internet stocks faltered. America Online was down 57% from its 52-week high by the end of 2000, and Amazon.com was down 82% from its 52-week high by the end of 2000. But the decline of these stocks did not diminish investor interest in the Internet. Instead, money shifted into what are perceived as the most attractive areas of technological innovation, including optical networking, with stocks like Nortel Networks (up 123% between November 1999 and August 2000 before dropping back to its November 1999 level) and Sycamore (up 150% from November 1999 to its 52-week high in March 2000 before dropping back 75% from its March high through November 2000) being the short-term beneficiaries.

These examples also illustrate how rapidly money flows in and out of sectors. Nortel's stock did very well until it announced a slowdown in the growth of its optical business. As the bellwether stock in the optical networking market, Nortel's announcement quickly soured investors on the sector.

The relationship between the performance of Internet stocks and that of their peers is very complex. When investors are afraid of falling behind, Internet stocks tend to go up the most because investors perceive that the Internet stocks are in the fastest-growing sector of the economy. When investors are afraid of losing their gains, they tend to sell the Internet stocks and invest the proceeds in DJIA stocks that are likely to hold their value better in a slowing economy. During these periods, the Internet stocks tend to perform less well than their peers.

DRIVERS OF INTERNET STOCK RELATIVE PERFORMANCE

Why do Internet stocks perform differently than other types of stocks? The interaction between the differing belief systems that we described in Chapter 1 is apparent in the flow of money into different kinds of investments. Money flows into Internet stocks and out of value stocks as momentum investors and venture-capital-oriented investors drive up the prices of Internet stocks so much that value investors feel compelled to sell their underperforming value stocks to invest in the Internet stocks. This money-flow pattern drives up the value of Internet stocks even more—causing the cycle to repeat itself.

Venture-capital-oriented investors believe that Internet stocks are transforming the economy—this perception of transformation is reinforced by profit gains at many companies. Furthermore, with almost half the families in the U.S. invested in direct contribution plans for their retirement, there is a continuous flow of fresh money into mutual fund families such as Janus, which made significant investments in Internet stocks. Finally, many investors believe that technology stocks in general are not as susceptible to efforts by the Federal Reserve to slow down the economy by raising interest rates.

An inescapable fact is that these money-flow cycles produced huge gaps in valuations between Old Economy and New Economy stocks—when traditional valuation metrics are applied to both sectors of the economy. Companies with strong long-term financial performance such as Fannie Mae, Emerson Electric, Berkshire Hathaway, Schering-Plough, and May Department Stores have been valued at much lower levels than technology companies, including such Internet infrastructure companies as Cisco Systems, Oracle, and EMC. These performance differences are summarized in Tables 3-1 and 3-2.

TABLE 3-1. SELECTED NEW ECONOMY STOCK MARKET VALUATIONS

Company	Price 4/4/00	Price 12/29/00	Price Change 9 Months	PE Ratio 1999	2000	Market Value (Bil.)	5-Yr. Profit Growth Rate*
Applied Materials	$98.75	$38.19	−61.3%	72.4	19.1	$35	25%
Cisco Systems	73.63	38.25	−48.1	203.2	89.1	212	30
EMC	64.00	66.50	3.9	137.5	97.6	141	30
Exodus Comm.	68.47	20.00	−70.8	NM	NM	7	65
JDS Uniphase	107.00	41.68	−61.0	NM	NM	42	45
Oracle	37.97	29.06	−23.5	121	26.3	175	25
Sun Microsystems	45.00	27.87	−38.1	116.0	45.5	94	20
Veritas Software	110.00	87.50	−20.5	NM	NM	31	50
Yahoo	168.00	30.06	−82.1	1,570.0	60.4	17	50

Source: Yahoo Finance
*Estimated

TABLE 3-2. SELECTED OLD ECONOMY STOCK MARKET VALUATIONS

Company	Price 4/4/00	Price 12/29/00	Price Change 9 Months	PE Ratio 1999	2000	Market Value (Bil.)	5-Yr. Profit Growth Rate*
AMR	$32.48	$39.19	20.7%	7.9	7.0	$6	8%
Berkshire Hathaway	55.6K	71.0K	27.7	52.1	45.5	102	14
Emerson Electric	55.38	78.81	42.3	17.3	22.9	32	11
Fannie Mae	58.50	86.75	48.3	13.8	19.4	78	13
May Dept. Stores	28.75	32.75	13.9	11.7	14.1	11	11
Philip Morris	22.31	44.00	97.2	7.1	11.6	97	12
Schering-Plough	40.63	56.75	39.7	27.4	33.1	78	16

Source: Yahoo Finance
*Estimated

In 2000, the valuation differences between New and Old Economy stocks converged somewhat. For example, the P/E ratio of the New Economy stocks in Table 3-1 declined from 370 in 1999 to 56 in 2000, while the Old Economy stocks sampled in Table 3-2 rose slightly, from 19 in 1999 to 21 in 2000. In the stock market, the New Economy stocks lost 45% while the Old Economy stocks surged 41%.

Money flows can help explain these differences. At one level of analysis, stock prices are driven simply by the supply and demand for shares. Individuals holding shares in value-oriented mutual funds—which invest in the New Economy stocks exemplified by those in Table 3-1—saw the value of their mutual fund shares plunge. After watching one too many news programs about dot-com disasters, these individuals got fed up with losing their money. As a consequence, they sold their shares in the growth fund. These sell orders forced the growth fund manager to go into the market to sell shares of the New Economy stocks to generate the cash needed to redeem the growth fund shares. The investors then put that cash in the most rapidly rising Old Economy stocks. The effect was to drive the New Economy stocks further down and to have the opposite effect on the Old Economy shares.

One underlying factor driving this movement of funds is that Old Economy and New Economy investors have different growth expectations. Many portfolio managers in the current stock market seek annual earnings gains of 20% to 30%. In the past, investors like Warren Buffett were satisfied with companies that increased profits by 15% a year. In fact, each of the stocks in Table 3-1 at least doubled in value between April 1999 and April 2000 but lost 45% of that increase between April and December 2000.

The problem came in 2000, when investors realized that large technology companies would not be able to sustain annual profit growth in the 20%-to-50% range. As Table 3-3 indicates, revenue growth at technology companies in the S&P 500 Index was 15.8% in 1999, following a 7.4% increase in 1998. The five-year average growth rate was 12.5%, and the average over the past 20 years has been about 10%. If technology stocks cannot maintain high growth rates in 2001, it is likely that investors will accord them even lower P/E ratios. In fact, technology sector earnings growth for the fourth quarter 2000 was 17%, declining to an anticipated 4% for the first quarter of 2001 (Meyer, 2001). As we will see in subsequent chapters, the process of creative destruction in the various sectors of the Internet business is likely to sustain these high growth rates for selected companies in selected industries. The key for successful Internet investing will be to discern these select few.

TABLE 3-3. S&P 500 TECHNOLOGY SECTOR GROWTH AND VALUATION, 1976–1999

Year	Companies in the S&P Tech Sector	Sales Growth Estimate	P/E Ratio	5-Yr. Profit
1976	21	13.9%	–	–
1977	21	14.3	13.92	–
1978	21	18.9	13.11	–
1979	20	16.1	12.15	–
1980	24	16.4	13.91	–
1981	30	12.3	12.6	17.1%
1982	32	9.3	16.75	16.9
1983	32	11.0	17.94	17.3
1984	36	16.4	15.04	17.6
1985	38	3.1	35.38	15.0
1986	29	6.9	21.81	13.5
1987	29	13.8	14.69	14.1
1988	32	15.5	13.50	13.2
1989	34	6.6	14.79	11.9
1990	34	7.6	12.31	12.2
1991	33	0.4	19.68	11.0
1992	33	5.7	22.15	12.6
1993	34	8.0	27.95	14.3
1994	36	13.4	19.74	15.9
1995	41	19.1	19.40	17.8
1996	49	9.0	27.26	19.1
1997	51	11.3	31.32	19.7
1998	54	7.4	53.51	20.3
1999	61	15.8	63.62	22.5

Source: *Barron's*

As anyone who has been following the media hype about the Internet realizes, there is more going on to drive the performance of Internet stocks than merely their relatively high rates of revenue growth. Beneath the surface of the statistics there are unique activities that drive the movement of money in and out of Internet stocks. Day traders, message boards, influential analysts, and the pervasive influence of the cable-TV network CNBC all have a real impact on the day-to-day flow of money in the markets. In some cases, these money drivers are simply new technologies that have speeded up the traditional process of sharing market gossip. In other cases, these phenomena are new and surprisingly powerful.

Day Trading

According to *Fortune*, day traders are an important driving force behind the high prices of many Internet stocks. For example, at Tradescape, a New York City–based day-trading firm, a 24-year-old trader named Adam Mesh was watching CNBC's Joe Kernen discuss a fuel-cell company called Plug Power (PLUG), which Mesh and his colleagues had never heard of (Schwartz, 2000). With 20 minutes left in the trading day, Mesh is watching PLUG on his computer screen. PLUG's stock price has already risen $15 to $53, then suddenly begins to climb again. Mesh buys 500 shares at $55. About a minute later, he gets out at $58. But PLUG is still soaring, so Mesh gets back in, this time at $60. Other traders in Tradescape's offices are also trading PLUG, and its shares move higher.

Without knowing what PLUG does, Mesh trades nearly 10,000 shares of PLUG, making $20,000 in the process. Mesh is far from the most successful trader in PLUG in those minutes. Another trader in the Tradescape offices announces that he has made $100,000 on PLUG.

Mesh's experience with PLUG is a microcosm of the changes that have been taking place in the composition of market participants. The indicator of these changes is the number of shares in a transaction. While institutional investors typically trade in blocks of 10,000 or 20,000 shares, individuals (including day traders) may buy or sell many fewer shares in a transaction, say 300—the average number of shares per trade in January 2000 of JDS Uniphase.

The smaller trade sizes indicate a significant increase in the proportion of market participants who are individuals instead of institutions. For example, since 1996 the size of the average trade on NASDAQ has dropped 50%, to just under 700 shares. Furthermore, the traditional buy-and-hold approach to investing has also become less popular. For example, the average NASDAQ stock is now held for just five months, down from two years in 1990. Trading volume has grown as well—January 2000 saw the highest volume ever for NASDAQ and the NYSE. These trends have worked together to increase volatility dramatically—in 10 of 21 trading sessions in January 2000, NASDAQ moved by 2% or more.

The overall impact on the market of individuals, including day traders like Tradescape clients, is significant. For example, on February 2, 2000, Tradescape traders accounted for just under 3%—40 million shares—of the NASDAQ's overall volume. Among technology companies that are particularly popular among Tradescape clients, this impact is even larger. For example, on that same day, Tradescape clients accounted for 10% of Qualcomm's volume and 15% of JDS Uniphase's (Schwartz, 2000).

Message Boards

Another new phenomenon driving the market is message boards. Message boards are online water coolers that connect market participants discussing topics that often relate to factors that could drive the behavior of a security. An important element of message boards is that the identity of participants is disguised. Therefore, it is impossible to know the financial interests of the participants or whether the content of their messages is accurate.

Even professional money managers, who are trained to regard day trading as little more than gambling, are forced to track the Yahoo chat boards and other online gossip centers that the amateurs frequent. Rod Berry, comanager of the RS Information Age fund, a top-performing technology fund, admits that the message boards are somewhat influential. While Berry does not trade based on information on the message boards, he does get ideas from them (Schwartz, 2000).

Occasionally, the day traders and the message boards can combine to exert a tremendous impact on the behavior of a particular security. For example, on February 2, 2000, Datron Systems, a maker of mobile communications technology, released information about what it claimed was a "breakthrough" that could allow high-speed Internet access from moving vehicles. The news release was placed on the wires at 8:21 A.M. When the stock began trading at 9:30 A.M., it was at $11^{15}/_{16}$, up $^{15}/_{16}$. At 10:01 A.M., the first message about the news arrived on Datron's Yahoo message board, and by 11 A.M., its shares had climbed to nearly $25. When trading in Datron ended for the day, after several mentions on CNBC, the stock was priced at 198, and 11.8 million shares had changed hands. Datron's average daily volume had been about 25,000.

The Datron case demonstrates the power of the message boards. Chris Hallahan, a 33-year-old from Sacramento, California, who began day trading in December 1999, saw the Datron news on the message boards and started buying. Hallahan saw a message about Datron at about 10 A.M. and bought about 15 minutes later. Hallahan looked at Datron's recent financial statements and some recent press releases on the Web and decided that the stock could reach $100 within a month. Hallahan noted that other companies in Datron's industry were valued at over a $1 billion, but Datron's market capitalization was a mere $50 million (Schwartz, 2000).

While the notion of passing rumors about a stock is nothing new, the ability to share those rumors on a platform that is open every minute of every day to investors all over the world is new and powerful. There is no doubt that message boards are susceptible to misuse by unscrupulous actors. This potential for misuse is particularly important because of cases like the movement of Datron's stock price on February 2, 2000. Investors in Internet stocks ignore the message boards at their peril.

Analysts

Although powerful securities analysts have had an influence on stock prices in the past, when these analysts began delivering their messages to millions of investors through new media such as the Internet and CNBC, novel effects were produced. In particular, a bold stock price prediction could gain an obscure analyst a huge amount of attention in a very short period of time. And this attention can influence the behavior of investors—driving the rapid movement of a stock's price.

Consider the case of Walt Piecyk, a 28-year-old telecommunications analyst with Paine Webber. Piecyk lacked the experience and the influential employer—Morgan Stanley, Goldman Sachs, Merrill Lynch—that traditionally leads an analyst to gain influence. On an otherwise quiet day in late December 1999, Piecyk contributed to a 31% rise in one stock, appeared on CNBC and in the *Wall Street Journal*, and was mentioned in

20 newspapers across the U.S. Piecyk achieved this by predicting that a wireless telecommunications firm, Qualcomm, then trading at $503 a share, could reach $1,000 within a year (Schwartz, 2000).

While predicting that a stock could double is not unusual, the media seem to become more interested when a stock's target price reaches into the four-figure range. Before the market opened at 9:30 A.M., Piecyk's prediction was repeated on CNBC and the Internet. By late afternoon, shares of Qualcomm had risen by $156 to $659.

While Piecyk had developed a 10-year earnings model for Qualcomm that helped him arrive at the $1,000 target, the real reason for the stock's performance was that an analyst had made a bold prediction, the media and the Internet had communicated this prediction, and retail investors had purchased the stock—probably without thinking much about whether or not the prediction was reasonable. Of the 60 million Qualcomm shares traded on this day, only one million were in institution-sized blocks. In retrospect, the institutions may have been wise not to purchase Qualcomm shares. By early February, shares of Qualcomm were down 15% from where they were the day after Piecyk made his $1,000-a-share prediction (Schwartz, 2000).

The foregoing discussion suggests that the prices of technology stocks in general, and Internet stocks in particular, are influenced by many forces that are difficult to understand and even more challenging to control. Competing belief systems, rapid dissemination of often inaccurate information, and ill-considered investment decisions by small investors all appear to play a role in the mechanism that determines where investment dollars flow and how prices are set. While the intent of this book is to sort out these complexities and identify a way to evaluate Internet investment opportunities, before diving into the details of the concepts introduced in Chapter 1, it is worth gaining the perspective of leading Internet industry analysts and their attempts to explain the behavior of Internet stocks.

VIEWS OF LEADING INDUSTRY ANALYSTS

In trying to make sense of the valuation of Internet stocks, analysts have approached the problem from opposite ends of the spectrum of tangibility. Some analysts start from the perspective that traditional methods of valuing stocks do not apply to Internet stocks. Based on this, these analysts invent a new set of measures for valuing Internet stocks. Other analysts believe traditional approaches to stock valuation do apply to Internet stocks. These analysts believe that Internet stocks are grossly overvalued based on traditional measures, using them to identify specific Internet businesses that are on the verge of running out of cash.

Both perspectives are helpful for our purposes here. An examination of the strengths and weaknesses of these approaches can help put the approach we introduced in Chapter 1 into some context. As we will explore now, the first approach is intriguing, but it seems to apply more to the e-commerce and Web portal segments than to Internet business segments that sell a more tangible product, such as the Web tools or Internet infrastructure segments. And while this first approach looks interesting, it remains somewhat difficult to use for predictive purposes. The second approach is quite useful for identifying stocks to avoid buying, but it is less useful in helping investors pick long-term winners.

New Accounting for the New Economy

One of the leading thinkers in the area of how to account for Internet companies is Baruch Lev, the Philip Bardes Professor of Accounting and Finance at New York University's Leonard N. Stern School of Business. According to *Fast Company*, Lev believes that accounting is increasingly irrelevant. Lev believes that the problem with traditional accounting is that it is over 500 years old. Luca Pacioli, an Italian mathematician who lived in Venice in the 1400s, developed double-entry bookkeeping so that managers could monitor their transactions and make sense of the way that they did business (Webber, 2000).

Today, argues Lev, value is created by intangible assets such as ideas, brands, ways of working, and franchises. Lev notes that the ratio between the market value of the Standard & Poor's 500 and their net-asset values is now greater than six. This means that the net-asset amount, as measured by traditional accounting on the balance sheet, represents only 10% to 15% of the value of these companies. This represents a substantial difference between value as perceived by those who pay for it day-to-day in the stock market and value as the company accounts for it.

He cites the parent of American Airlines, AMR, as illustrating the massive amount of value inherent in these intangible assets. He notes that in October 1996 AMR sold 18% of its computer-reservations system, SABRE, to the public. AMR retained the remaining 82%. In January 2000, SABRE constituted 50% of AMR's value. Lev was amazed that one of the largest airlines in the world, with 700 jets in its fleet, 100,000 employees, and exclusive landing rights in heavily trafficked airports, was valued by the stock market as much as SABRE, a computer reservations system.

He highlights an important difference between the ability to get additional business value out of tangible versus intangible assets. He notes that the ability to get additional business value out of tangible assets is limited, but the ability to get additional value out of intangible assets is virtually unlimited. To illustrate the value of tangible assets, he notes that it is impossible to use the same airplane on five different routes at the same time. By contrast, he notes that SABRE works as well with five million people as it does with one million people. Lev notes that knowledge assets, such as SABRE, are subject to increasing returns to scale, where the larger the network of users, the greater the benefit to all network participants.

Lev notes that Pacioli's method does not keep adequate track of any of these intangible assets. The failure of traditional accounting to monitor intangible assets affects financial analysts, corporate financial officers, employees, and managers. This disconnect makes it hard for employees to value their contributions to the company and challenges managers trying to decide whether to back a project or to assess a project's performance. Lev argues that traditional accounting and financial reporting methods do not help assess whether knowledge-based companies are overvalued on the stock market or whether companies are paying too much to acquire knowledge-based assets.

Lev provides useful examples of how accounting practices are at odds with the business practices that they are trying to measure. One such example comes from America Online (AOL). In 1994 and 1995, AOL counted part of its customer-acquisition costs as assets. In accounting parlance, AOL capitalized these costs. By using this accounting policy, AOL was saying that, in acquiring new customers, it was creating an asset that would help AOL become even more profitable in the future. Lev points out that financial analysts called AOL's accounting practices inappropriate. Since AOL was in a new industry where competition was fierce, analysts thought that AOL was trying to manipulate its earnings. In October 1996, AOL gave up and completely expensed its $385 million in customer-acquisition costs. In April 2000, AOL had a market value of roughly $140 billion. Lev notes that in comparison, the $385 million that AOL tried to capitalize—and the related analyst outrage—was so small as to be trivial in comparison to the value that the customer-acquisition costs created.

He provides another example—IBM acquisition of Lotus in 1995—to highlight the way traditional accounting requirements nullify corporate accountability. As an accounting requirement, IBM had to estimate the fair-market value of the assets that it had acquired. IBM estimated that the portion of Lotus's R&D that was in process—R&D for which there was not yet a product—was worth $1.84 billion. That amount represented 53% of the $3.5 billion acquisition price. According to accounting rules, once in-process R&D is estimated, it must be expensed. IBM followed the accounting rules and thereby eliminated the R&D asset that it purchased from Lotus. Lev points out that this kind of mindless writing off of all investment in knowledge assets eliminates accountability and the ability to measure the performance of an investment or to learn from it.

Lev is developing a Knowledge Capital Scoreboard, which attempts to quantify intangible assets. His method measures knowledge assets, intellectual earnings, and knowledge earnings. His computation starts with "normalized earnings," a measure based on past *and* future earnings. In accounting for knowledge, Lev believes it is essential to consider the potential for future earnings that knowledge creates.

He looks at the past and the consensus forecasts of analysts. Based on those forecasts, he creates an average, which he calls "average normalized earnings." From those normalized earnings he subtracts an average return on physical and financial assets, based on the theory that these are substitutable assets. Merck, for example, has laboratories and manufacturing facilities. While he notes that the equipment there is not unique, the people, patents, and knowledge that is being developed there are unique. After he subtracts from the total normalized earnings a reasonable return on the physical and financial assets, Lev defines what remains as knowledge earnings. Those are the earnings that are created by the knowledge assets.

Lev gave several examples of knowledge assets based on his January 2000 calculations. He estimated that Microsoft had knowledge assets worth $211 billion—by far the most of any company. Intel's knowledge assets totaled $170 billion, and Merck's were worth $110 billion. While DuPont had more employees than all of those companies combined, Lev calculated DuPont's knowledge assets as worth $41 billion.

He noted that different kinds of knowledge assets make a significant difference. He calculated that Philip Morris had knowledge assets worth $160 billion, largely because of its brand values. Coca-Cola is also a significant brand, thus its knowledge assets were worth $60 billion. Lev also identified another type of knowledge asset—structural capital, defined as a unique way of doing business. Dell Computer does not produce computers that are necessarily better than other companies' computers, but the way in which Dell markets its computers is entirely different. This difference in structural capital explains why Dell's knowledge capital totals $86 billion—higher than that of Wal-Mart.

He believes that to complement the knowledge assets measure, he must also identify the drivers of knowledge. Lev suggests that companies that end up surviving the Darwinian process have superior innovation skills. To measure these skills quantitatively, he is developing a technological-capabilities index based on measures of inputs, intermediate outputs, competitive position, and ultimate output. The input measures include investment in R&D, product development, and information systems. The measures of intermediate outputs include items such as patents and trademarks. The measures of competitive position include the number of people who access a particular Web site. Finally, Lev is developing measures based on the ultimate output—commercialization. His overall objective is to develop a system of knowledge and innovation drivers—a system that allows managers to measure things that work or do not work, both of which indicate a company's knowledge assets (Webber, 2000).

While Lev's efforts appear promising, it is important to note that traditional accounting still has enormous power, particularly when it can help investors assess the most basic aspect of a company's operation—its ability to generate enough cash to pay its bills. It is in this area that the work of a *Barron's* reporter, Jack Willoughby, attracted significant attention in March 2000.

Burn Rate Accounting

Burn rate is a jarring ceiling above which Internet hype cannot rise. The collision between the hype associated with the Internet and the rate at which a particular Internet company consumes its available cash creates terror—the kind of terror that contributed to a 13% decline in the NAS-DAQ within a few hours on April 3, 2000. The foot that started the ball rolling on this terrifying drop in value was an article published a few weeks before in *Barron's*. The article predicted that the Internet bubble would soon burst for many companies that would run out of cash by the end of 2000 unless they were able to issue new debt or equity to stave off the grim reaper (Willoughby, 2000).

This study made investors realize that Internet companies were not totally immune from the traditional limits of economics. The *Barron's* conclusion was based on a study by Pegasus Research International, which indicated that at least 51 Internet firms would "burn through" their cash by the end of 2000. This amounted to a quarter of the 207 companies included in the study. Firms predicted to run out of funds most quickly were CDNow, Secure Computing, drkoop.com, Medscape, Infonautics, Intraware, and Peapod. Table 3-4 illustrates how this analysis was applied to the companies with the fewest months until burnout as of March 2000.

TABLE 3-4. MONTHS TILL BURNOUT FOR 10 INTERNET COMPANIES, MARCH 2000

| | Results for Calendar 4th Qtr. '99 | | | | |
Company	Market Value 4/6/00	Revenues	Operating Expenses	Operating Losses	Months Till Burnout
Pilot Network Services	$400.2	$8.47	$3.83	$–4.62	–0.14
CDNow	118.9	53.11	44.34	–34.69	0.37
Secure Computing	293.9	6.90	11.43	–7.60	1.51
Peapod	54.50	21.56	14.55	–9.07	2.01
VerticalNet	4,040.0	10.09	17.10	–10.18	2.20
MarketWatch.com	415.6	10.00	28.70	–22.46	2.52
drkoop.com	81.5	5.10	25.73	–20.63	3.34
Infonautics	82.5	5.85	7.93	–2.07	3.41
Medscape	228.9	4.02	27.99	–23.97	3.42
Intelligent Life	31.3	3.75	17.04	–13.29	3.51

Source: Pegasus Research International, Yahoo Finance
Note: Dollars in millions.

The Pegasus study made some assumptions that were considered controversial at the time. For example, the study assumed that the rates of loss for the fourth quarter of 1999 would be annualized and that the companies would not be able to raise additional funds. Based on these assumptions, 74% had negative cash flows. For many, there seemed to be little hope of near-term profits. While many of the companies in the study were small, one of the best-known companies, Amazon.com, was indicated to have only 10 months' worth of cash remaining—before it managed to raise $690 million by issuing convertible bonds in early 2000. Even that cash was predicted to last Amazon only 21 months.

While some of the firms in the most imminent danger were able to find acquirers or raise additional cash, other Internet firms were less fortunate. For example, Peapod, a Web-enabled grocery delivery service, announced in March 2000 that its chairman, Bill Malloy, intended to resign. As a result, investors who had agreed to provide $120 million in financing were backing out. Peapod hired an investment bank to seek new investors, but with $3 million of cash on hand, Peapod was not expected to survive the month. Fortunately for Peapod, Dutch grocer Royal Ahold injected sufficient capital into Peapod for it to continue operating as Stop & Shop's online grocer. As noted earlier, CDNow was acquired by Bertelsmann.

In 1998 and 1999, investors were willing to bid up the shares of these easy-to-understand business-to-consumer companies. The high stock prices made profitability irrelevant because the companies could easily issue new shares at higher prices to raise the cash they needed. An example is eToys, a toy retailer that came public at $20 and rose above $80. While the public understood the concept, the competition, including Toys R Us, launched their own Web sites, and investor enthusiasm for eToys diminished. By March 2000, eToys shares sank to 11¾. Investors who purchased shares of eToys between $20 and $80 a share were unlikely to purchase more shares, even at $12. *Barron's* estimated that eToys had enough cash to last until February 2001 (Willoughby, 2000).

The counterpoint of Jack Willoughby's cash burn rate accounting with Professor Lev's Knowledge Capital Scoreboard highlights an important challenge facing investors. Willoughby's use of traditional accounting numbers to assess how much cash is left before burnout is a very useful analysis for investors looking for stocks to avoid. Lev's Knowledge Capital Scoreboard could be helpful as a way of identifying which stocks to buy. The juxtaposition of the two methods highlights the ongoing value of traditional accounting as well as its limits. As we will develop throughout the remainder of this book, the most useful guide for investors in Internet stocks is a blend of the old and the new.

INTERNET STOCK INVESTMENT PRINCIPLES

Successful investing in Internet stocks demands that investors understand all the dimensions of a company—from the profit potential of its industry to a specific company's ability to generate sufficient cash to fund its operations. The Internet Investment Dashboard we presented in Chapter 1 is based on these principles and forms the basis of our analysis in subsequent chapters. We will be applying these principles to each of the nine Internet business segments—showing how to gather the information required to make intelligent Internet investment choices.

Internet Investment Principle 1. Invest in industries with economy bargaining power. Look for companies that participate in industries where overall returns are likely to increase. Pay close attention to revenue growth rates, profit margin trends, market share distribution, changes in the number of competitors, and the height of entry barriers in the company's market. While this sounds pretty intimidating, the good news for investors is that most of this information is available in the company's prospectus or 10K— both documents are available at EDGAR (www.sec.gov). There are sections on "business," "market," "competition," "strategy," and "technology" that generally provide a sober analysis of these factors. If an investor conducts this analysis for a company and its competitors, useful insights can emerge about whether any company in the industry is ever likely to earn a profit, and if so, how much of a profit and how sustainable that profit is likely to be. Investors should read these sections closely because many Internet companies target "land-based" markets in which it is not yet clear whether an online opportunity can be exploited profitably.

Internet Investment Principle 2. Invest in companies with the potential to sustain market leadership. Find companies that lead their industry and whose leadership position is growing. Look at market share fluctuations, changes in the level of customer satisfaction with the company, rates of employee turnover and hiring, rate of new product introduction, and changes in analysts' views of the company. To track these factors, follow developments in industry publications such as *Business 2.0, CNET, ZDNet, The Industry Standard, Red Herring, Upside, Fortune,* and *Business Week Online.* In addition, there are many analyst companies that follow the Internet, including Forrester Research, Gartner Group, Jupiter Media Metrix, and Aberdeen Group. The Internet companies themselves are also good sources of information about their relative market position—often their Web sites will reveal useful information as well. For example, Cisco Systems offers the results of Internet infrastructure market analysis reports that show Cisco as the leader in a number of different product categories.

Internet Investment Principle 3. Invest in management teams that adapt effectively to change. Look for management teams that anticipate change and act, not just react. Examine a company's acquisition rate; how effective it has been at integrating those new concerns into the company; and the stream of new strategic alliances. Analyze a firm's method of monitoring customer feedback, new competitors, and new technologies. Using the same sources mentioned above, investors can follow the developments of the industry and how a company is adapting. For example, Cisco Systems has been very adept at acquisitions, having made at least 55, and still counting, since 1993. Following the evolution of the industry, it is clear that Cisco is trying to keep up with changing customer needs and the ever more powerful technologies designed to do a better job of meeting those needs.

Internet Investment Principle 4. Invest in companies that brand themselves memorably. Seek out firms that graft their personalities onto the minds of investors and customers. Check out the track records of the firm's venture capital investors, investment bankers, management team, partners, and customers. Some of these factors may be surprising to an individual investor, but they are important. Certain venture capitalists—such as Kleiner Perkins, Sequoia, Benchmark, Draper Fisher Jurvetson—are perceived as better than the rest. These tier-one players are highly regarded because they have invested in well-known winners in the past, including Apple Computer, Cisco Systems, Yahoo, and Amazon. While many investors do not understand the technology that Internet companies are using, they can understand the language of enormous investment returns. Investors figure that if a tier-one venture capitalist has invested in a particular company, it is sending a signal of value that this company is likely to be a winner.

Internet Investment Principle 5. Invest in companies that make effective use of financial resources. Investors should seek out firms that manage their resources effectively. Investors should track items like marketing expenses per new customer, research and development expenses as a percent of new product revenues, gross profits per customer, revenues per employee, and cash burn rate.

Internet Investment Principle 6. Invest in companies whose market values are reasonable in comparison to peers. Timing of investment decisions is important, particularly in the short run. Investors should seek out companies that fit well with the previous five principles and whose stocks are undervalued relative to their peers. Key indicators here are market-capitalization-to-sales ratios, revenue-growth-to-market-capitalization-to-sales ratios, market-share-to-market-capitalization-to-sales ratios, and price/earnings ratios to earnings growth rate (for profitable Internet industry segments).

CONCLUSION

Internet stocks behave differently than do Old Economy stocks. While the factors that drive their market price performance are unique, this does not mean that investors cannot apply discipline to the process of buying and selling Internet stocks. Rather, Internet stocks demand a unique discipline that blends many of the elements of traditional securities analysis—such as financial effectiveness and relative market valuation—with a more in-depth understanding of industry dynamics, competitive positioning, management effectiveness, and branding.

Chapter 4

_____

WEALTHY TRAFFIC COP: INTERNET INFRASTRUCTURE STOCK PERFORMANCE DRIVERS

W HILE SOME MAY FIND the topic of Internet infrastructure dull and technical, those who care about earning high investment returns cannot help but find the topic enthralling. Simply put, the investment returns in Internet infrastructure have been and are likely to continue to be among the highest available. As we will see, the reasons for these high returns pertain to the ability of Internet infrastructure firms to sell a product that enhances the speed and performance of the Internet itself. No matter who wins the e-commerce battles, all participants need the ability to direct Internet traffic—hence the counterintuitive notion of the wealthy traffic cop.

The field of Internet infrastructure is undergoing a fundamental transformation. The leaders of the new world are enjoying significant stock price appreciation, leaving behind the leaders of the old world.

Nortel is an intriguing example of an old-world firm that has transformed itself into a new-world winner. For example, between September 1999 and September 2000, Nortel's stock appreciated almost 250%. The former Northern Telecom is a Canadian company whose sports-car-loving CEO, John Roth, made a sharp left-hand turn in 1995 that has made all the difference. Through the acquisition of Bay Networks in 1998, Nortel was reenergized through the infusion of Bay Networks' faster-moving management style. But the key to the success of Nortel's operations is its $10 billion worth of optical-equipment sales, which make Nortel the dominant player in the latest wave of technology to accelerate the transmission of information over the Internet.

By contrast, Cisco Systems, which has dominated the world of enterprise routing since it began operations in the mid-1980s, has lost some of the momentum in the stock market. Between September 1999 and September 2000, Cisco's stock price increased about 75%. Of more concern for investors, Cisco's stock has crumbled since it peaked in March 2000. The company is fighting a two-front war—attempting to maintain its dominant position in enterprise routing while building a strong share of the much larger telecommunications-equipment market. While Cisco continues to make acquisitions, including in the area of optical networking, it lags behind Nortel in optical networking, even as new routing competitors—such as Juniper Networks—take market share in Cisco's core router market. At the same time, Cisco is losing some of its management talent, most notably Cisco CEO heir-apparent Don Listwin, who left in the summer of 2000 to head up Phone.com, an IP telephony firm.

While the general public has a dim understanding of Internet infrastructure, it is by far the most important Internet-related industry. As we noted in Chapter 2, every dollar's worth of e-commerce revenue generates $1.49 worth of Internet infrastructure revenue. The importance of this dimly understood industry is highlighted by the fact that for a brief period in April 2000, the leading company in the Internet infrastructure segment, Cisco Systems, surpassed Microsoft and General Electric to become the company with the highest market capitalization in the world.

INTERNET INFRASTRUCTURE
STOCK PERFORMANCE

In this chapter we begin with an examination of the stock price performance of the Internet infrastructure segment. By Internet infrastructure we mean the systems that direct traffic over networks—whether they be enterprise or service-provider networks. By enterprise, we mean all organizations that are not Internet service providers or other traffic carriers. The systems that direct traffic are generally specialized computer systems called routers and switches.

The essential function of these devices is to move traffic from its source to its destination. The problem of directing such traffic is very complex for a number of reasons. First, different organizations along the path that the traffic will travel may use different communications protocols—or languages—for transmitting the data. As a consequence, routers must act like massive translation tables that can connect traffic among protocols. Another challenge for routers is analyzing alternative paths for transmitting data from source to destination and identifying which of those paths is the most efficient at a particular moment.

Given the exceptionally rapid growth of network traffic, to which we alluded earlier in this book, it is apparent that the router is a critical element of a network's operation. To the extent that a vendor of routers can develop new technologies that accelerate the speed with which traffic can be routed at a lower cost, these new technology providers can anticipate significant spikes in demand. The substitution of these technologies drives the relative stock price performance of participants in the Internet infrastructure industry.

We noted in Chapter 3 that the stock market performance of a cross section of Internet infrastructure firms was 144% in 1999. While this performance was strong compared to the NASDAQ's 86% increase in 1999, the 144% increase did not compare favorably to the performance of the nine Internet business segments on average, which increased 339%. However, in 2000, the Internet infrastructure firms lost 36% of the 1999 gain, a relatively strong performance compared to the 67% decline of the Internet index and the 39% drop in the NASDAQ.

As Figure 4-1 illustrates, the stock market performance of Internet infrastructure firms varied quite widely. During 1999, the relative performance of stocks in the industry was related largely to the market's perception of their future prospects more than their past performance. For example, while Cisco Systems has sustained its market leadership in routers and grown to become the largest company in terms of market capitalization, its stock price performance within the sector was about average.

By contrast, Juniper Networks' stock value increased almost twice as much as Cisco's despite the fact that Juniper Networks had a financial track record of losing money and a relatively limited market share. Juniper Networks' stock benefited from investors' perceptions that its future growth potential was greater than Cisco's—partially due to the widespread acceptance of Juniper Networks' relatively fast router by a number of leading network service providers.

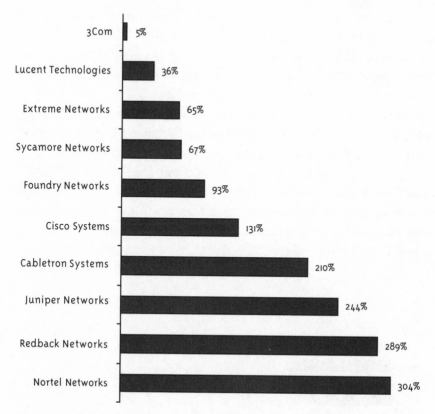

Figure 4-1. Selected Internet Infrastructure Firms' Percent Change in Stock Price, 12/31/98–12/31/99

Source: MSN MoneyCentral

Note: Foundry Networks, Juniper Networks, Redback Networks, Extreme Networks, and Sycamore Networks percentages are calculated from the closing price on the first day of their 1999 initial public offerings to 12/31/99.

In fact, by the third quarter of 2000, Jupiter had 30% of the market for core routers, a market which Cisco had traditionally dominated.

The stock market performance of the Internet infrastructure firms was substantially different in 2000. As Figure 4-2 indicates, one firm that investors perceived as a market leader performed exceptionally well in the stock market even as many other Internet stocks took a beating. The top infrastructure stock market performer in 2000 was Juniper Networks, which truly distinguished itself in its market as we will see later in this chapter.

By contrast, several of the weaker Internet infrastructure firms lost significant stock market value during 2000. In particular, Foundry Networks, Lucent Technologies, and 3Com all lost a substantial amount of stock market value during 2000. As we will explore later in this chapter, the stock market performance of these firms was related to their product market position and their resulting financial performance. For example, investors punished Lucent for missing its profit targets.

The decline in 2000's Internet infrastructure stock market performance was particularly severe after Nortel's October 2000 announcement of a slow-down in optical networks equipment growth. The November 2000 bank-ruptcy of ICG Communications, a Competitive Local Exchange Carrier (CLEC), also contributed to investor fears, since CLECs were big customers for this network infrastructure. Analysts' forecasts that telecommunications equipment spending might decline in 2001 caused investors to fear slowing growth in the Internet infrastructure market.

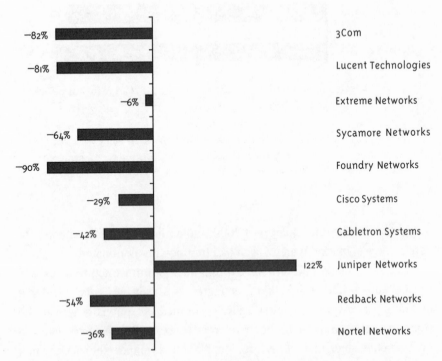

Figure 4-2. Selected Internet Infrastructure Firms' Percent Change in Stock Price, 12/31/99—12/29/00

Source: MSN MoneyCentral

INTERNET INVESTMENT DASHBOARD ANALYSIS OF INTERNET INFRASTRUCTURE

While the foregoing discussion of the relative stock price performance of Internet infrastructure segments is suggestive, it does not offer a fully satisfactory explanation. To obtain such an explanation, we apply the Internet Investment Dashboard (IID) analysis to the Internet infrastructure segment.

As we will see, the IID analysis suggests that Internet infrastructure is likely to generate attractive investor returns in the future. The Internet infrastructure industry clearly has economic bargaining power, as illustrated by its attractive profit margins. Selected participants that offer customers a closed-loop solution are likely to sustain the greatest share of industry profits. Several industry participants have demonstrated their ability to adapt effectively to change and have created compelling brand families. Many participants manage their financial resources effectively, and a handful are priced relatively cheaply.

Economic Bargaining Power

The market for Internet infrastructure has significant economic bargaining power. The market is large, and participants earn attractive profit margins. For internet infrastructure providers profit margins vary by relative market share. Firms with the highest relative market share tend to have higher margins because they are able to sell additional products with lower incremental sales costs. More fundamentally, the reason firms can generate attractive margins is that their products are important to powerful decision-makers. If a network does not process information fast enough, a company ceases to operate at full effectiveness. Companies depend on their networks to operate and are therefore willing to pay what they need to in order to assure that operation.

The complexity of the technical challenge that Internet infrastructure attempts to solve is mirrored in the complexity of analyzing its market. Internet infrastructure consists of a variety of different submarkets composed of the different groups of customers and different technologies. Different analysts analyze the Internet infrastructure market differently, thereby making it difficult to provide a single clear picture of the Internet infrastructure market. When analyzing the revenues of Internet infrastructure firms such as those highlighted in Figure 4-1, it is useful to understand that while some of the firms generate all their revenues from just one of the submarkets, other firms derive revenues from all or a combination of the submarkets.

For large firms like Cisco Systems, market size matters. Market size matters for large firms because the only way for large firms to maintain high revenue growth is to take a significant and growing share of larger markets—particularly after they have dominated and saturated the markets in which they initially competed. For example, Cisco Systems controls roughly 80% of the market for routers purchased by companies (as opposed to network service providers). As a result, Cisco Systems is looking toward the much larger telecommunications-equipment market in order to generate continued 30% to 50% revenue growth as it exceeds $20 billion in annual revenues.

For smaller firms like Juniper Networks, the relevant issue is rapid growth. A small firm in a rapidly growing market can sustain very high revenue growth merely by keeping up with the market's average growth rate. If that small firm is particularly skilled, it can expect to grow revenues at an even faster rate. As we will explore later, the stock market appears to reward investors in firms whose revenues grow the fastest with faster stock price growth rate than their slower-selling peers.

The largest Internet infrastructure market is telecommunications equipment. According to a September 1999 research report from Sutro & Co., the total worldwide market for telecommunications equipment is about $200 billion a year. This market includes five types of telecommunications products:

- **Switches.** The worldwide market for switching equipment, which includes local exchange switches and tandem switches, reached about $40 billion a year in 1999, and each segment is growing by about 5% to 10% annually.
- **Wireless equipment.** Sales of wireless equipment, which includes wireless switches, are currently at $50 billion and are growing by 20% to 25% a year.
- **Transport and transmission equipment.** The global market for transport and transmission equipment is $20 billion and is growing between 25% and 30% annually.
- **Voice-processing equipment.** Sales reached about $20 billion in 1999.
- **Data-communications equipment.** Driven by an increase in data traffic

of about 30% annually, this market reached $40 billion a year in 1999 and is growing about 15% annually (*Cambridge Telecom Report*, 1999).

While the telecommunications-equipment market is larger, it is not growing as fast as the optical-networking-equipment market. Through 2001, the growth in spending on telecommunications equipment is likely to be capped by hostile capital markets and a slowing economy. In the meantime, the formerly fast-growing enterprise-network-equipment market is not only smaller, but is also growing more modestly. According to another market researcher, The Dell'Oro Group, the rapid growth of the Internet has caused a significant market shift. The previously fast-growing market for corporate data networking equipment is slowing, while the core of the high-speed Internet, optical networking, is anticipated to grow fastest. More specifically, The Dell'Oro Group predicted 69% growth for optical-networking-equipment sales in the year 2000, compared with 15% growth for corporate data-networking equipment such as hubs, switches, and routers. Another market research firm, Ryan, Hankin & Kent (RHK), estimates that the annual worldwide optical-networking market will reach $14 billion by 2002 (Masud, 2000).

Another category of the Internet infrastructure market that is driving very high stock market valuations is the gigabit/terabit switch and router market. These devices are very-high-speed mailing machines. The terms gigabit and terabit refer to the number of pieces of information that these machines are capable of processing each second. Gigabit means that the machines can process billions of pieces of data—or bits—per second, and terabit means that they can process trillions of bits per second.

While different analysts have different forecasts, there appears to be a broad consensus that the gigabit/terabit switch and router market is likely to grow rapidly and become significant within the next three years. As Figure 4-3 suggests, the core switching and router market is anticipated to reach $5.5 billion by 2003, growing at a 100% compound annual rate from 1998's $169 million in revenues (Masud, 2000).

The Internet Infrastructure industry consists of many segments that vary in size and growth rate. While it is difficult to pinpoint precisely how large each segment is, there are clear trends emerging in terms of future opportunity for industry participants. First, the most rapid growth appears likely to take place in emerging technologies, such as optical networking and gigabit/terabit switches and routers. These new technologies are antic-

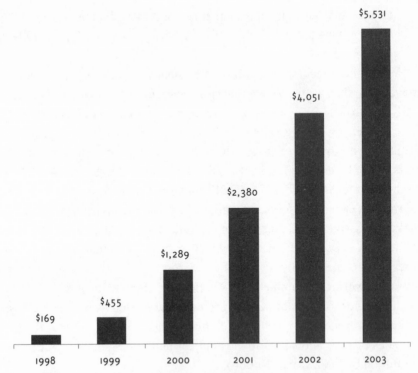

Figure 4-3. Core Switch and Router Market Size Forecasts, 1998–2003,
in Millions of Dollars

Source: RHK

ipated to grow the most rapidly because they promise to relieve congestion at the core of the Internet—thereby accelerating the realization of the Internet as a high-speed, multimedia network. Second, more traditional corporate-networking equipment is anticipated to grow more slowly. Third, large network-equipment providers will need to gain market share in larger, more rapidly growing markets in order to sustain high revenue growth. Whether this rapid revenue growth can ultimately be converted into higher profit growth remains to be seen.

While the question of future profit growth remains unknown, the margins in the Internet infrastructure business are relatively high. As Table 4-1 indicates, the mean gross margin of 10 Internet infrastructure firms is 60%. This relatively high gross margin suggests that the average participant in the industry is able to set a sufficiently high price to offset its cost of goods sold. The 11% net margin for the nine companies (excluding Redback) indicates that the average firm in this sector is actually earning positive net income.

TABLE 4-1. MARGINS FOR 10 INTERNET INFRASTRUCTURE FIRMS, DECEMBER 2000

Company	Stock % Change	Gross Margin	Operating Margin	Net Margin	2-Year Trend
Nortel	304%	45%	−2%	−6%	−
Redback	289	65	−380	−380	——
Juniper	244	67	32	21	++
Cabletron	210	50	43	25	−
Cisco	131	71	22	14	−
Foundry	93	72	35	25	+
Sycamore	67	67	4	0	0
Extreme	65	64	9	6	+
Lucent	36	49	12	7	+
3Com	5	47	17	10	+
MEAN	**146%**	**60%**	**19%***	**11%***	**0**

Source: MSN MoneyCentral

Note: Two-year trends represent the extent to which net margins have changed in the most recent two years. ++ signifies a large improvement, + indicates a small improvement, 0 indicates no change, − indicates small deterioration, —— indicates large deterioration. Stock % change is for 1999.

*Excludes Redback

Table 4-1 also indicates that there is no strong correlation between profit margins and stock price performance. To the extent that the table suggests any relationships between stock price performance and margins, there appears to be a slight inverse correlation between the trend in profit margins and stock price performance. Simply put, the stock prices of firms whose profit margins were not improving seemed to go up more than those whose margins were improving.

While the amount of information here is not sufficient to draw definitive conclusions, the table suggests that the stock market is rewarding firms that are deferring short-term profit growth and instead investing in acquisitions and R&D in order to take advantage of large future market opportunities. Perhaps the stock market perceives that the firms that focus principally on margin improvement are harvesting the remnants of a once-promising business and lack exciting opportunities for future investment.

The Internet infrastructure segment has economic bargaining power.

The segment produces revenues in the hundreds of billions of dollars and is growing anywhere between 3% and 100%, depending on the product category. This scale and growth suggest that the value of these products is sufficiently high to encourage powerful decision-makers around the world to write big checks to purchase them. Furthermore, the profit margins that leading companies are able to generate suggest that purchasers of the products—whether companies or network service providers—are sufficiently dependent on the continued survival of Internet infrastructure suppliers that they are willing to allow them to earn average net margins of 11%.

Closed-Loop Solution

While it is clear that the Internet infrastructure market is relatively attractive, the next question to analyze is which industry participants offer customers the best value. The concept of the closed-loop solution is particularly important to buyers of network equipment because the typical network consists of many different pieces of equipment made up of different technologies from different vendors. Buyers of network equipment prefer to purchase new equipment that will work with their existing network. In addition, buyers prefer to purchase network equipment that will increase the speed of their network and enhance its security and reliability, all at a lower cost.

Vendors who satisfy all these needs are likely to offer customers the highest return on their investment in Internet infrastructure equipment. And offering this high return depends on the firms' relative ability to knit all the pieces of the network fabric into a relatively seamless Web—a closed-loop solution. The challenge of the Internet investment analyst is to devise ways to quantify this concept, or at least to gather objective data to assess its existence.

One way to assess which firms are offering customers a closed-loop solution is to determine these firms' relative market shares. The extent to which a company is offering customers a closed-loop solution can also be measured to a certain extent by its relative sales levels, sales growth rates, and market capitalization. Finally, it is possible to analyze the extent to which the firms are performing all the activities in their business that would be required to provide customers with a closed-loop solution.

The data on relative market shares suggest that Cisco Systems is the dominant player in the Internet infrastructure business. Table 4-2 compares the relative market share and revenue growth rates of the top five firms in the multiservice wide area network (WAN) switch and WAN router product categories. Figure 4-4 shows the share of revenues for the five leading participants in a third product category, edge switches and routers.

What becomes clear from examining the data is that Cisco Systems is the dominant player in these three product categories. But closer inspec-

TABLE 4-2. MARKET RANK AND REVENUE GROWTH OF TOP FIVE PARTICIPANTS IN THE MULTISERVICE WAN SWITCH, Q4 1999, AND WAN ROUTER MARKETS, Q3 1999

Vendor	Multiservice WAN Switch Market Rank	Multiservice WAN Switch Sales Growth	WAN Router Market Rank	WAN Router Sales Growth
Cisco	1	14%	1	47%
Lucent	2	26	3	−53
Nortel	3	45	4	0
Newbridge	4	22	NA	NA
Juniper	5	54	2	68
Ericsson	NA	NA	5	50

Source: The Dell'Oro Group

Note: Multiservice WAN data are for the fourth quarter of 1999. WAN Router data are for the third quarter of 1999. Cisco held 78% of the WAN router market, while Juniper controlled 15%.

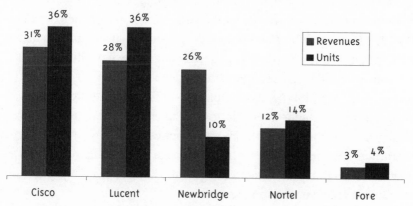

Figure 4-4. Edge Switch and Router Revenue and Unit Share of Top Five Vendors, First Six Months of 1999

Source: RHK

tion reveals that Cisco Systems faces considerable challenges to that leadership. For example, in the multiservice WAN switch category (estimated 1999 revenues of $5.7 billion), Lucent and Nortel are growing much more quickly than Cisco—relative growth that could constitute a challenge to Cisco's leadership. In the WAN router category (estimated 1999 revenues of $800 million), Juniper's 68% growth also could challenge Cisco. And in edge switching and routing (an estimated $3 billion market), Cisco is vying for leadership with Lucent.

This analysis of relative market share suggests that market-share momentum is an important factor. For example, Juniper Networks experienced far greater stock price appreciation than Cisco Systems, and Juniper's revenues grew much faster as well. As a result of its tie with Cisco in share of units in this market, Lucent might well have experienced much higher price appreciation if it had not missed its fourth-quarter 1999 earnings forecast.

Another way of assessing the relative effectiveness of the strategies of firms in the Internet infrastructure market is to analyze the firms' relative size, revenue growth, and market capitalization. This analysis can help us gain additional insights into which firms are generating the greatest product market momentum and the extent to which that momentum has translated into greater stock price appreciation.

The correlation between momentum in the product markets and the capital markets is partially born out by the evidence. As Table 4-3 indicates, in some cases the higher the rate of revenue growth, the higher the firm's rate of shareholder value creation. The stock market seems to reward acceleration in revenue growth for the larger firms, having increased Nortel's 1999 stock price by 304% largely because of its successful acquisition strategy that put the company in a stronger competitive position.

Conversely, the market punishes large players like 3Com that are shrinking in a rapidly growing market—reflecting the stock market's belief that management is unable to implement a strategy that creates a leading market position in the more attractive segments of the Internet infrastructure business. Smaller firms like Redback and Juniper, which are growing very rapidly in attractive segments, have enjoyed rapid rates of shareholder value appreciation as well.

TABLE 4-3. REVENUE MOMENTUM OF 10 INTERNET INFRASTRUCTURE FIRMS, DECEMBER 2000, IN MILLIONS OF DOLLARS

Company	Stock % Change	12 Months Revenue	Quarterly Revenue Growth	Market Capitalization
Nortel	304%	$28,450	41%	$96,600
Redback	289	190	328	5,420
Juniper	244	424	561	37,200
Cabletron	210	1,290	−24	3,130
Cisco	131	21,530	68	267,200
Foundry	93	327	247	1,340
Sycamore	67	299	515	9,540
Extreme	65	334	152	4,430
Lucent	36	39,450	−9	57,000
3Com	5	4,588	−33	3,520

Source: MSN MoneyCentral

Note: Stock % change for 1999, market capitalization as of January 10, 2001.

The stock market rewards Internet infrastructure firms that are establishing leadership positions in attractive market segments. While the specific technology market segments that investors perceive as attractive are likely to continue to change, the stock market seems to reward the firms that either develop leading technology themselves or are capable of acquiring that technology and integrating it into their own operations to provide customers with a closed-loop solution. Conversely, firms that are not able to position themselves to win in attractive market segments are often punished with a very slow rate of shareholder value growth.

Management Integrity and Adaptability

Management integrity and adaptability are far more difficult to measure than a firm's relative market share. Nevertheless, investors who can make astute judgments regarding management integrity and adaptability are likely to avoid stocks that are likely to decline in price and find the long-term winners. Integrity is important because customers, employees, suppliers, and shareholders are risking time and money on the ability of managers to make commitments and stick to them. Management teams that can set ambitious goals and achieve them build layers of trust that encourage long-term relationships between the firm and the best customers, employees, suppliers, and shareholders. Building and motivating the best stakeholders enables the firm to outperform its competitors.

Financial statements provide clues about management integrity. For example, in 2000 there were examples of firms that were forced to restate their revenues as a result of accounting policies that were perceived as a bit too "aggressive." One example of this was a software company called MicroStrategy, which provides tools for analyzing and identifying meaningful patterns in large volumes of data such as consumer purchasing patterns. MicroStrategy was forced to change its revenue accounting policy in a way that cut its reported revenues and, hence, profits, leading to a drop in stock price and the filing of shareholder lawsuits.

Management adaptability is also critical because in Internet infrastructure, as in all Internet business segments, the rate of change is very rapid. Technology evolves rapidly, creating products that offer higher levels of performance at lower costs. Customer requirements typically become more demanding over time. New competitors emerge, often from unexpected places, and present significant challenges to an incumbent's market share. As a result of this rapid rate of change, investors must have a way to assess how well the management teams of the companies in which they invest can adapt to this change. Managers who can anticipate important changes and reposition their firms to profit from change are likely to sustain excellent financial performance. The ability to attain consistent financial performance in the context of rapidly changing product markets is a good indicator of management adaptability.

As Table 4-4 indicates, three of nine leading Internet infrastructure firms have legal proceedings against them, creating concern about their managements' integrity. Cabletron Systems has been sued for alleged improper reporting of revenues and was subject to a Securities and Exchange Commission (SEC) inquiry regarding its practices of accounting for acquisition-related charges. Cabletron's CEO has since been replaced. Lucent Technologies is involved in lawsuits regarding its financial responsibilities for cleaning up Superfund sites and has been sued for allegedly misrepresenting its financial condition in the fall of 1999, leading to a negative earnings surprise that hurt its stock. 3Com is involved in a number of legal proceedings, including five securities class-action suits, one patent litigation with Xerox, and a contract dispute with Disney Interactive.

TABLE 4-4. REVENUE ACCOUNTING POLICY AND SHAREHOLDER LAWSUITS FOR SELECTED INTERNET INFRASTRUCTURE COMPANIES

Company	Revenue Accounting Policy	Legal Proceedings
Redback	Recognizes product revenues, net of allowances, when it ships products to customers, if there are no outstanding vendor obligations and collection is considered probable.	None
Juniper	Recognizes product revenue at the time of shipment, assuming that collectibility is probable, unless it has future obligations for network interoperability or has to obtain customer acceptance, in which case revenue is deferred until these obligations are met.	None
Cabletron	Recognizes revenue on shipment of products and software. In addition, the SEC forced Cabletron to restate its accounting for certain acquisition-related charges.	Nine 1997 lawsuits challenging accounting practices[1]
Cisco	Generally recognizes product revenue on shipment of product unless there are significant postdelivery obligations or collection is not considered probable at the time of sale. When significant post-delivery obligations exist, revenue is deferred until fulfilled.	None
Foundry	Recognizes product revenue on shipment to customers.	None
Sycamore	Recognizes revenue from product sales on shipment if a purchase order has been received or a contract has been executed, there is no uncertainty regarding customer acceptance, the fee is fixed and is likely to be collected. If there are uncertainties regarding customer acceptance, revenue is recognized when uncertainties are resolved.	Immaterial lawsuits
Extreme	Generally recognizes product revenue at the time of shipment, unless Extreme has future obligations for installation or has to obtain customer acceptance, in which case revenue is deferred until these obligations are met.	None

Company	Revenue Accounting Policy	Legal Proceedings
Lucent	Revenue from product sales of hardware and software is recognized at time of delivery and acceptance and after consideration of all the terms and conditions of the customer contract.	"Superfund" litigation and 12 class action lawsuits alleging misrepresentation of financial condition
3Com	Generally recognizes a sale when the product has been shipped, risk of loss has passed to the customer, and collection of the resulting receivable is probable. Accrues related product return reserves, warranty, other postcontract support obligations, and royalty expenses at the time of sale.	Five securities class-action suits, one intellectual property suit with Xerox, and one breach of contract suit with Disney Interactive

1. Nine 1997 lawsuits were consolidated into one 1998 complaint, which alleges that Cabletron's alleged accounting practices resulted in the disclosure of materially misleading financial results during part of 1997. More specifically, the complaint challenged Cabletron's revenue-recognition policies, accounting for product returns, and the validity of certain sales.
Source: Company Financial Statements (10Ks)

The stock market seems to treat each of these cases differently. Cabletron's stock was rewarded when its board replaced senior management and announced a restructuring plan in which it intends to sell its various subsidiaries. Lucent Technologies was punished in the stock market for missing its earnings forecasts. By October 2000, Lucent's board decided to replace CEO Rich McGinn with Henry Schacht, Lucent's 66-year-old former chairman. 3Com's legal problems have been overshadowed by its strategic problems, resulting in basically flat stock price performance. In each case, investors have responded to questions regarding management integrity by selling the stock and have rewarded or punished the firms based on their responses to perceived breaches.

In light of the rapid rate at which the competitive environment changes, the ability of managers to adapt to this change effectively is a crucial factor in determining the long-term value of an Internet company. To analyze this factor, we can examine the variance of the percentage change in quarterly earnings. Simply put, if an Internet infrastructure firm is able to increase profits fairly consistently in each quarter, its management is good at adapting to change. Conversely, if a firm's quarterly performance jumps around significantly, its management team is probably less effective at adapting to change. The important point here is that the ability to adapt effectively to change in a rapidly changing environment can be measured by management's ability to produce fairly consistent financial results.

This measure is particularly useful for analyzing firms that have been publicly traded for a relatively long period of time. As Figure 4-5 indicates, the firms that have generated the lowest variance in annual percentage change in earnings per share seem to be the ones whose stock prices have performed better. Conversely, 3Com, the firm with the highest variance in annual percentage change in EPS, had the worst stock market performance.

The ability to manage acquisitions successfully has been the primary factor in differentiating the firms with the steadiest increases in earnings per share from the ones that have had the choppiest performance. Simply put, firms like Cisco Systems and Nortel have done a good job of making

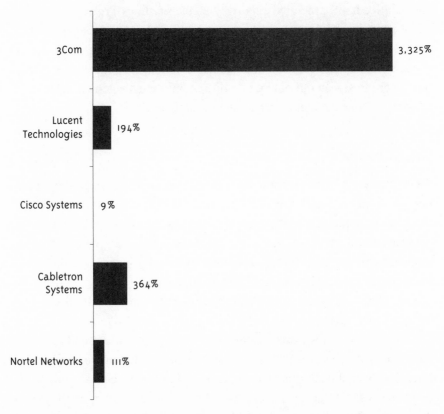

Figure 4-5. Variance in Annual Earnings Percent Change for Selected Internet Infrastructure Firms, 1995–1999

Source: MSN MoneyCentral, Peter S. Cohan & Associates Analysis

the right acquisitions and integrating them, and firms like 3Com have made many acquisitions that made limited sense strategically and were integrated poorly. For example, one of the primary reasons that 3Com encountered a significant earnings decline in 1998 was the botched acquisition of modem maker U.S. Robotics. As part of the integration of this acquisition, 3Com managed to produce far too much inventory, which it subsequently wrote off. By contrast, Cisco Systems has made over 55 acquisitions as of April 2000 and has turned the process of acquiring and integrating companies into a science that seems to contribute to the company's incredibly steady increase in annual earnings.

Management integrity and adaptability are critical factors in differentiating winning Internet companies from their peers. Firms, such as Cisco Systems, that have demonstrated consistently high levels of integrity and adaptability have generated relatively stable earnings growth over time, while firms like Cabletron Systems have suffered from management that lacked the same level of integrity and adaptability—leading to its current process of dismemberment. Internet investors can use the metrics illustrated here to screen out potentially disappointing management teams and to screen in management teams that are likely to excel.

Brand Family

Since it is often difficult for investors to understand the details of Internet infrastructure technology, investors can analyze the firms' brand families as a gauge of the firms' relative value. The conclusion of the analysis of the different firms' brand families is that there are some important differences among the firms, particularly in terms of their customers and principal investors. These differences foreshadow different shareholder-value-generation capabilities and are therefore useful indicators for investors.

An analysis of the firms' customer lists suggests that some of the firms have done a much better job than others at penetrating the most influential reference accounts. Firms that depend heavily on resellers seem to be rewarded less well by investors than firms that depend primarily on a direct sales force. Part of this problem is that resellers demand a piece of the revenue stream of relatively low-margin products, whereas a direct sales force can establish more lasting business relationships with large organizations. These lasting business relationships constitute a cost of changing vendors, or switching cost, that leaves more room for higher profit margins.

As Table 4-5 indicates, the firms with the best customer lists are such creators of shareholder value as Lucent, Nortel, Cisco, and Juniper. Conversely, the firms that have weaker customer lists and that are more dependent on resellers for their revenues, such as Cabletron, Extreme Networks, and 3Com, have not performed as well in the stock market.

Similarly, the firms with the most influential investors have generated

TABLE 4-5. MAJOR CUSTOMERS OF SELECTED INTERNET INFRASTRUCTURE COMPANIES, 1999

Company	Major Customers
Redback	UUNET (a subsidiary of MCI WorldCom), SBC, Southwestern Bell Information Services, and Pacific Bell Internet have been, since inception, Redback's largest customers in terms of revenues. Other customers include Ameritech, Bell Canada, Bell South, Concentric, EarthLink, Flashcom, Korea Telecom, Verio, and @Work, a division of @Home.
Juniper	UUNET (MCI WorldCom), Cable & Wireless, AT&T/IBM Global Services, Frontier GlobalCenter, and Verio.
Cabletron	During the year ended February 28, 1999, the company had one customer, Compaq, that accounted for 11% of total sales. End-user customers include brokerage, investment banking firms, and other financial institutions; federal, state, and local government agencies; commercial, industrial, and manufacturing companies; multinational and international companies; telecommunications companies; internet service providers; health-care facilities; insurance companies; universities; and leading accounting and law firms.
Cisco	Cisco Systems has customers in enterprise, small-business, telecommunications and service provider, and consumer markets. No customer accounts for more than 10% of sales.
Foundry	In 1999, sales to Mitsui accounted for less than 10%, and sales to Hewlett-Packard and America Online accounted for 14% and 11% of our revenue, respectively.
Sycamore	Sycamore has a small number of customers. Williams Communications accounts for substantially all of Sycamore's revenues.

Company	Major Customers
Extreme	In fiscal 1999, Compaq and Hitachi Cable accounted for 21% and 13%, respectively, of our net revenue. Compaq is both an OEM and an end-user customer. Customers include Amoco, AT&T, Hewlett-Packard, PSINet, Barnes and Noble, Honeywell, Qwest, British Telecom, Raytheon Cable & Wireless (UK), Real Networks, Lockheed Martin, Reuters, Compaq, Microsoft, Schlumberger, Dell Computer, Shell Oil, Sun Microsystems.
Lucent	Customers include AT&T, SPN's largest customer, as well as other large service providers such as the Regional Bell Operating Companies. Lucent Enterprise segment services a million customer locations worldwide that range in size from small to medium to large businesses. A significant amount of the Enterprise segment's revenue comes from approximately 250 global accounts.
3Com	For the year ended May 28, 1999, Ingram Micro and Tech Data Corporation accounted for 17% and 12% of 3Com total sales, respectively. 3Com customers include individual consumers of personal electronics, large global corporations, and communications services providers. Business enterprise customers include companies in a wide variety of industries, including finance, health care, manufacturing, telecommunications, government, education, and retail.

Source: Company Financial Statements (10Ks)

greater shareholder value than the firms with less prestigious backers. Firms whose largest holders are the most successful venture capital firms, such as Sequoia (one of the original investors in Cisco Systems and Apple Computer) and Kleiner Perkins (one of the original investors in Netscape, Yahoo, and other success stories), tend to generate better investor returns than established firms whose shares are held largely by institutional investors and, to a lesser extent, by senior management.

As Table 4-6 indicates, the firms with the most prestigious investors are such creators of shareholder value as Redback, Juniper, and Foundry Networks. Cabletron is an exception to this general pattern. Its major investors are not distinguished venture capital firms, and the primary source of Cabletron's significant stock price increase date is the market's estimation of the value of Cabletron's assets once they have been parceled out and sold to different groups of investors following the departure of its former CEO, who was responsible for driving Cabletron's price so low. As a general rule, Internet infrastructure firms, including Lucent and 3Com, whose major investors are large institutions, tend to perform less well. Generally speaking, high-performing venture capital investors tend to earn the highest returns from Internet infrastructure firms. With the exception of Cisco Systems, once firms become owned primarily by institutional and public investors, their share prices seem to perform less well.

TABLE 4-6. MAJOR INVESTORS IN SELECTED INTERNET INFRASTRUCTURE COMPANIES

Company	Major Investors
Redback	Sequoia Capital, members of senior management
Juniper	Kleiner Perkins, Ericsson Business Networks, Fidelity Management and Research (FMR), members of senior management
Cabletron	ex-CEO
Cisco	FMR, Barclays Bank, Janus Capital, members of senior management
Foundry	Crosspoint Venture Partners, Institutional Venture Partners, Accel Partners, Vantage Venture Partners, members of senior management
Sycamore	Matrix Management, Platyko Partners, members of senior management
Extreme	AVI Capital, Kleiner Perkins, Norwest Equity Partners, Trinity Ventures, members of senior management
Lucent	Barclays Bank, FMR, State Street, members of senior management
3Com	Capital Research and Management, Barclays Bank, T. Rowe Price, members of senior management

Source: Company Proxy Statements

While investors may wish to conduct other brand family analyses, including partnerships and management team backgrounds, the point should be clear from the foregoing examples: The better the brand, the greater the shareholder returns. The key concept here is that many investors simply do not understand the technology and business models of Internet infrastructure firms. As a result, they tend to look for brand-name stakeholders who are willing to lend their reputations to the firms they believe are most likely to enhance their already-strong reputations by winning in the marketplace. Clearly this principle applies to the Internet infrastructure market, where the best creators of shareholder value have the strongest brand families, and the weaker-performing firms in the stock market have weaker brand families.

Financial Effectiveness

An important indicator for investors in Internet infrastructure stocks is how well the various firms in the industry are using their financial resources. The industry includes a mix of firms, including some that are growing profitably at a moderate pace and others that are growing revenues very quickly, albeit unprofitably. To compare these firms in terms of financial effectiveness, it makes sense to look at their sales productivity and overhead ratios.

There is a wide variation among the firms in the industry in terms of their sales productivity. As Figure 4-6 suggests, the productivity figures should be analyzed in two groups—those of startup firms and those of larger firms. Among the startup firms with the highest sales productivity are Juniper, Foundry, and Extreme. Their products are in such great demand that they do not require a large sales force—the performance of the products themselves constitutes the primary criterion on which sales are closed. Among the larger firms, Cisco Systems has by far the most productive people. Cisco's sales per employee are about twice as large as those of its large-company competitors. It is worth emphasizing, however, that Cisco Systems cannot maintain its market leadership or sustain its stock price performance over the long term if it is not able to close the sales productivity gap with the startup competitors.

Internet infrastructure firms also seem to fall into two camps as pertains to their efficiency in spending not-product-specific resources. Simply put, the larger, more established firms seem to spend less of their revenue on sales, general, and administrative expenses than do the upstart competitors. In general, the larger firms have already established their sales forces and are not adding to them as quickly as the small firms. To a certain extent, investors seem to reward firms that spend a larger proportion of their revenues on SG&A, interpreting this spending as an investment in establishing market leadership. Certainly, Redback and Juniper are among the stocks that have performed best while spending more on SG&A than their upstart firm peers.

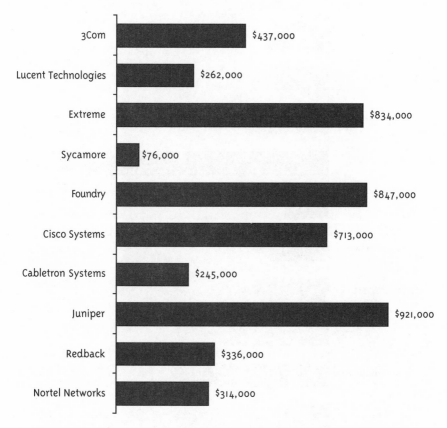

Figure 4-6. Sales per Employee of Selected Internet Infrastructure Firms, 1999
Source: MSN MoneyCentral

Two smaller firms do not seem to follow this pattern. As Figure 4-7 indicates, Sycamore Networks spent more than its revenues on SG&A and was in the middle of the pack in terms of stock price appreciation. Foundry was a superior performer in the stock market but spent a relatively small amount of money on SG&A. Similarly, the most efficient of all the firms in terms of SG&A as a percent of sales was Nortel, which had the best increase in stock market value in 1999.

The conclusion from this analysis is that investors should understand measures of financial effectiveness in the broader context of the firms' strategic evolution. More specifically, firms that have established market leadership in relatively mature market segments may be expected to spend less money on SG&A and have lower sales productivity than their upstart competitors. In general, the market seems to reward firms that have the

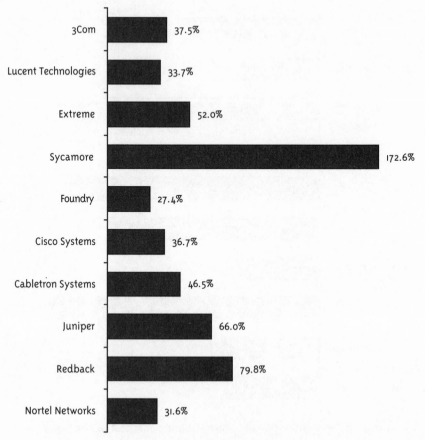

Figure 4-7. Selling, General, and Administrative Expense as a Percent of Sales for Selected Internet Infrastructure Firms, 1999
Source: MSN MoneyCentral

highest sales productivity—the upstarts—and expects these rapidly growing firms to spend heavily on SG&A to establish market leadership in rapidly growing market segments.

It also makes sense for investors to understand the relative levels of financial effectiveness for companies within a peer group. For example, over a longer period of time, the market seems to reward large firms with relatively high sales productivity, such as Cisco Systems, and to punish those with lower sales productivity, such as Cabletron. As we noted earlier, this principle was reversed in 1999 as investors rewarded Cabletron with a greater increase in share price because of its turnaround from a losing firm going nowhere to a play on the salvage value of its various business units.

Relative Market Valuation

While the foregoing factors are all important for the evaluation of Internet infrastructure stocks, they are only precursors to the most fundamental measure—the relative value of the firms from the perspective of the securities market. The fundamental question facing investors is whether a stock that is trading at a price that represents a relatively low ratio of sales or book value is more or less likely to go up in value. Simply put, investors need a way to determine whether the market is marking down a particular stock because it is less likely to go up in value or more likely to go up in value.

Having examined the five other gauges on the Internet Investment Dashboard, we are now in a better position to answer this question. As Table 4-7 indicates, the market seems to be assigning widely disparate values to the sales of the leading Internet infrastructure firms. One interesting point is that the stock market seems to place the highest value on each dollar of cash flow, the second highest value on each dollar of sales, and the

TABLE 4-7. STOCK PRICE AND REVENUE GROWTH PERCENT CHANGE, 2000, AND VALUATION RATIOS FOR 10 INTERNET INFRASTRUCTURE FIRMS, JANUARY 10, 2001

Company	Stock % Change	Revenue Growth	Price/ Sales	(Price/Sales)/ Rev. Growth	Price/ Book	Price/Cash Flow
Nortel	−36%	41%	3.45	8.41	4.39	42.4
Redback	−54	328	25.81	7.87	1.10	−160.0
Juniper	122	561	86.83	15.48	61.75	329.9
Cabletron	−42	−24	2.17	−9.04	2.08	6.7
Cisco	−29	68	12.2	17.94	9.52	63.0
Foundry	−90	247	46.1	1.68	4.88	15.2
Sycamore	−64	515	33.07	6.42	6.19	158.2
Extreme	−6	152	11.09	7.30	8.45	86.0
Lucent	−81	−9	1.54	−17.11	1.99	14.8
3Com	−82	−33	0.85	−2.58	0.80	4.30
MEAN	**−36%**	**185%**	**18.12**	**3.64**	**10.12**	**56.05**

Source: MSN MoneyCentral

least value on each dollar of shareholders' equity. These differences may suggest some useful insights for investors. In particular, when looking at revenue growth and price/cash flow valuations, it appears that some of the fastest-growing firms are among the most valuable.

Typically, such firms are the most cash-flow-starved because they invest most heavily in building a solid position in rapidly growing markets. As a result, price/cash flow ratios are highest for the most valuable firms—possibly because the market anticipates that these firms will earn a significant share of the fastest-growing market segments and hence eventually be able to generate substantial cash flows. A consequence of achieving this anticipated outcome would presumably be a drop in the price/cash flow level as the firm achieved greater revenues and higher levels of cash flow.

Table 4-7 also suggests the importance of interpreting valuations based on making a distinction between incumbent and upstart firms. Simply put, the typical incumbent firm has a much lower average valuation than an upstart. The reason for this difference is that the incumbent firms grow much more slowly than the startups. But the variations in valuations among incumbent firms highlight potential investment opportunities. For example, Nortel and Lucent may be significantly undervalued relative to Cisco. While Cisco may well be a much better managed firm, it is not clear that Cisco is so much better managed that it warrants a price/sales ratio that is almost six times that of Lucent and Nortel.

While the valuations of the upstart firms are much higher, the price/sales ratios do appear to correlate roughly with their relative rates of sales growth. The upstarts' high valuations are most vulnerable to variations in their revenue growth rates. Simply put, if firms like Juniper or Redback experience significant slowing in their growth rates, their valuations are likely to drop as well. Needless to say, as these firms achieve significant scale, it will be more difficult—if not impossible—for them to achieve the triple-digit revenue growth rates that they have been achieving to date.

The Internet Investment Dashboard analysis suggests that Internet Infrastructure could be an attractive segment for long-term investors. In Figure 4-8, two of the indicators on the dashboard are clear, while four have a dotted pattern. In particular, the potential for slowing telecom equipment spending in 2001 could lower revenue growth and profit margins for the industry. These indicators suggest that investors should carefully distinguish among individual companies within the sector. Firms like 3Com and

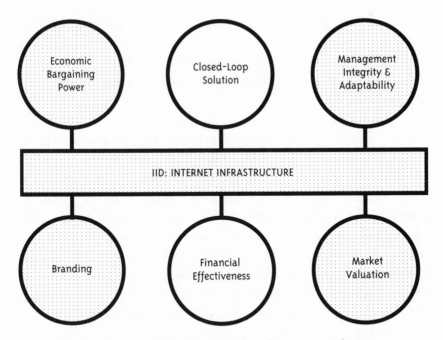

Figure 4-8. Internet Investment Dashboard Assessment of Internet
Infrastructure Segment

Key: Clear = positive performance and prospects; dotted = mixed performance and prospects.

Cabletron may be worth avoiding unless there is some significant change in the six factors we have analyzed. Firms like Nortel and Lucent may be worth considering based on their relative valuations. Upstarts like Juniper Networks may warrant investment if they can continue to gain market share and/or if there are significant market dips that could create interesting buying opportunities. Cisco Systems is a special case because its persistent market leadership has made it one of the most popular holdings in the world. As a result, Cisco Systems could be a good long-term investment to purchase during market corrections—assuming that Cisco Systems continues to perform well on the six indicators outlined in this chapter.

PRINCIPLES FOR INVESTING IN INTERNET INFRASTRUCTURE STOCKS

Needless to say, the analysis we have conducted here is subject to change. Undoubtedly some of the strong-performing companies will falter, some of the weak ones could generate positive surprises, and new competitors will emerge to tilt the playing field in a totally different direction. As a result, investors in Internet infrastructure stocks need some principles to guide them through the analytical process as investment conditions evolve.

Internet Infrastructure Investment Principle 1. Monitor developments in new technologies that can enhance bandwidth and security. Investors must recognize that the position of the Internet infrastructure industry is most likely to remain attractive. The most important factor that will determine the sector's long-term profitability is the emergence of new technologies that offer customers wider bandwidth with greater security at lower costs. Investors who shun old technologies and seek out new ones that offer greater customer value are likely to achieve superior returns. In addition, investors should monitor the financial health of telecom companies, a key customer group, to help assess the profit potential of Internet infrastructure.

Internet Infrastructure Investment Principle 2. Analyze the evolution of the upstarts' strategies to assess which will be able to offer closed-loop solutions. Investors should recognize that incumbent firms are more likely to offer customers closed-loop solutions. Investors should assess the quality of these solutions among incumbents and pick the specific firm that offers the most effective closed-loop solution. At this writing, Cisco Systems is the market leader. Over time, however, this could change if firms like Juniper Networks are able to integrate new technologies into their product lines and to offer customers a superior value proposition. The most successful investors in Internet Infrastructure will monitor this evolution closely and shun firms that are unable to offer closed-loop solutions to customers or whose ability to offer these solutions erodes.

Internet Infrastructure Investment Principle 3. Avoid investing in firms where there are questions about management integrity and adaptability. Our analysis indicated questions about management integrity and adaptability in a few firms. This analysis is most useful for investors as a sell signal. If a firm with a track record of avoiding shareholder lawsuits and generating consistent profit growth suddenly deviates from this path, investors should consider selling the stock. In many cases, such deviations from the path of integrity and adaptability could be a signal that management is no longer up to the job. When these deviations occur, it is likely that management will try to gloss over the problems. Investors may not want to stay along for the ride.

Internet Infrastructure Investment Principle 4. Use compelling brand families as a signal of value in conjunction with other indicators. Even in the absence of a long financial track record, firms like Juniper Networks have such compelling brand families that there is a reasonable chance that they will be able to build a significant business. As new Internet infrastructure firms are taken public, it is likely that there will be some excellent investment opportunities for the investor who can shrewdly assess the quality of their brand families. The best ones may warrant investment, and the worst ones may be worth avoiding.

Internet Infrastructure Investment Principle 5. Employ financial effectiveness measures in the context of firms' stages of development. For many of the established Internet infrastructure firms, an assessment of financial effectiveness can be a useful tipoff to management problems. If a firm's financial effectiveness is eroding, as measured by a trend of growing SG&A/sales or sales/productivity ratios, this may suggest that the time to sell has arrived. For upstart firms, however, high levels of SG&A/sales may go hand in hand with high sales productivity. Here it may be more important to monitor deterioration in the sales-productivity figure relative to peers in order to identify specific firms that may be faltering.

Internet Infrastructure Investment Principle 6. Monitor relative valuations most closely within strategic groups to identify potentially under- and overvalued companies. While we reject the notion that every company has an intrinsic value to which its market capitalization can be compared, it is clear that comparing market valuations of peer firms can highlight individual companies that the market may be unfairly punishing or rewarding. These mismatches among peer groups can create opportunities for investors who know how to look for them. Since market conditions can change so frequently, this last principle is one that investors should monitor more frequently than some of the others mentioned above.

CONCLUSION

Internet infrastructure is an attractive area for investors. Because the industry has economic leverage and because there are many firms with good strategies, excellent management, and compelling brand families, it is up to the investor to assess the six indicators—particularly, the industry's economic bargaining power—to identify the individual firms within the industry that are likely to prevail. Valuation matters for this segment. Depending on the investment time horizon, investors may want to watch closely for opportunities to purchase shares in segment market leaders during periods of market decline in which excellent companies are punished along with poor ones. Such well-timed investments could yield attractive returns.

Chapter 5

VOLATILE FUEL FOR THE FIRE: INTERNET VENTURE CAPITAL STOCK PERFORMANCE DRIVERS

INTERNET VENTURE CAPITAL is an area full of paradoxes. While investors and owners of many privately held venture capital firms have created huge fortunes through many astute Internet investments, the owners of shares in publicly traded Internet venture capital firms have gained and lost—depending on the timing of their buys and sells. While the scrutiny that goes into investing in individual Internet companies may be very high, it is extremely difficult for an investor in publicly traded Internet venture capital firms to gain true insight into the value of these firms' individual portfolios. As a result, investors who can make direct investments in privately held Internet companies may profit, while investors in publicly traded Internet investment companies may be in for a wild ride.

Benchmark Capital is a private venture capital firm that had the good fortune of earning the single highest return on investment in the history of venture capital. This return was achieved by investing $2 million in an online auction site, eBay, in June 1997. This investment peaked at $2.3 billion following eBay's IPO, yielding Benchmark a return of 114,900%. Although eBay never touched the initial $2 million, eBay used Benchmark to help recruit to its staff toy industry experts and executives from Pepsi. While this credibility certainly helped eBay get off the ground, it remains unclear how much Benchmark contributed to its 114,900% return on investment and how much was sheer luck. Certainly, Benchmark's other investments—particularly those in the B2C area—did not perform nearly as well. Nevertheless, the investors in Benchmark's private funds are happier than the public shareholders of Internet Capital Group.

Public investors in firms such as Internet Capital Group (ICGE) have not necessarily earned such attractive returns. While those who purchased ICGE shares soon after its IPO in August 1999 saw the stock rise from $23 to $212 in December 1999, 1999's meteoric rise was decimated as ICGE plunged 98% to $3.28 by the end of 2000. The ICGE story highlights the real dangers of investing on the basis of a compelling concept whose underlying reality is difficult to quantify. In the case of ICGE, the compelling concept was the notion of buying into the potential returns of CMGI—a B2C investment star during the late 1990s—for an even bigger opportunity area—B2B. Simply put, investors thought that buying ICGE during the fall of 1999 would allow them to get in on the ground floor of the next big thing—only 10 times bigger. Unfortunately, the April 2000 Internet stock crash changed that mind-set, leaving ICGE investors holding stock in a collection of impossible-to-evaluate, privately held B2B companies with limited prospects for profits.

Internet venture capital is the fuel that has driven the growth of Internet business. As we will soon see, the level of venture capital directed toward the various sectors of the Internet business has grown dramatically. Furthermore, the exceptionally high returns that venture capital firms enjoyed as a result of the 1999 strength in the IPO market attracted even greater levels of venture capital from a variety of sources. These sources include traditional limited-partnership investments in private venture capital firms, corporate venture funds, angel investors, and a few publicly traded venture capital firms. Nevertheless, the April 2000 dot-com crash marked the beginning of a downturn in VC returns. By the second quarter of 2000, VC returns dropped to a paltry 4%, and as more VC-backed dot-coms such as pets.com went out of business, VC returns were expected to go negative.

INTERNET VENTURE CAPITAL STOCK PERFORMANCE

Currently, there are very few publicly traded Internet venture capital firms. The first well-known publicly traded Internet venture capital firm was CMGI, whose shares have been publicly traded for several years. In 1999, Internet Capital Group was taken public. Its stock increased in value quite dramatically in 1999, only to fall back substantially during the year 2000. U.S. investors may have some difficulty investing in and tracking other publicly traded Internet venture capital firms, such as Japan's Softbank or Hong Kong's Pacific Century Cyberworks.

As Figure 5-1 illustrates, the stock market performance of the two publicly traded U.S. Internet venture capital firms was very strong in 1999 and deteriorated sharply throughout 2000. As we will see throughout this chapter, the performance of these companies' stocks suggests two important challenges facing investors in publicly traded Internet venture capital firms. The most significant challenge for investors is that the financial statements of Internet venture capital firms do not provide a significant amount of insight into the value of the firms' portfolio companies or of their executives' investment acumen. The second challenge facing investors is that the stocks of these publicly traded Internet venture capital

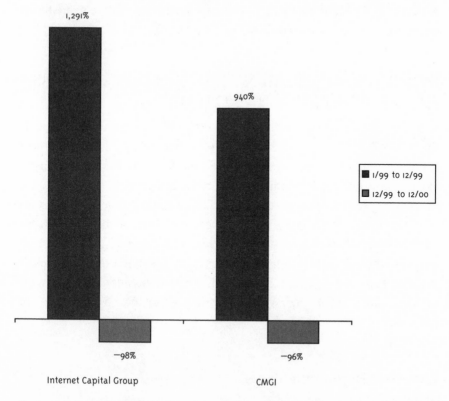

Figure 5-1. Publicly Traded Internet Venture Capital Firms' Percent Change in Stock Price, 12/31/98–12/31/99 and 12/31/99–12/29/00.

Source: Yahoo Finance

firms tend to swing up and down based on the momentum belief system that we outlined in Chapter 1. More specifically, investors tend to chase these stocks when the sectors in which CMGI and Internet Capital Group invest are rising in popularity. And investors tend to chase the stocks down when the sector is declining in popularity.

Simply put, publicly traded Internet venture capital firms are extremely sensitive to transient investor sentiment. CMGI did well in the stock market when business-to-consumer (B2C) e-commerce was popular. Internet Capital Group did well from August to December 1999, when business-to-business (B2B) e-commerce was popular. When the popularity of both B2C and B2B plummeted in the spring of 2000, the shares of CMGI and ICGE followed suit. The nature of these firms is that some of their portfolio companies can be valued based on publicly traded companies in the same industry, but others are virtually impossible to assess with any certainty. As a result, investors in publicly traded Internet venture capital firms will need to make judgments about expected future value based on the investment track records of the firms' managers.

INTERNET INVESTMENT DASHBOARD ANALYSIS OF INTERNET VENTURE CAPITAL

The IID analysis suggests that the Internet venture capital segment is likely to remain attractive over the long term, depending on the investment acumen of the firms' management. The Internet venture capital industry clearly has economic bargaining power, as illustrated by its very attractive investment returns. Most participants offer customers a closed-loop solution, and these participants are likely to gain the greatest share of industry profits. Several industry participants have demonstrated their ability to adapt effectively to change and have created compelling brand families. It is difficult to assess how effectively they use their financial resources, however, and whether their valuations are high or low. We now explore each of these factors in turn.

Economic Bargaining Power

The market for Internet venture capital is large and participants earn attractive returns. Significant economic bargaining power is enjoyed by Internet venture capital firms with the following attributes: significant capital; access to excellent managers, partners, and customers; strong ties to prestigious underwriters; and significant negotiating power over portfolio companies. Simply put, the best Internet venture capital firms consistently gain access to the best deals and can negotiate the most favorable terms with portfolio companies. For portfolio companies, there is a huge benefit to gaining investment from one of the top venture capital firms.

Venture capital has grown very dramatically over the last seven years. As Figure 5-2 indicates, the total amount of venture capital raised grew over 10-fold, from about $4 billion in 1993 to $46 billion in 1999. While the growth rate of venture capital raised has accelerated in recent years, it is perhaps most noteworthy that in 1999 the amount of venture capital raised exceeded the amount of buyout funding for the first time since 1985. This development is suggested in Figure 5-3.

This crossing of the lines between venture capital and buyout capital in

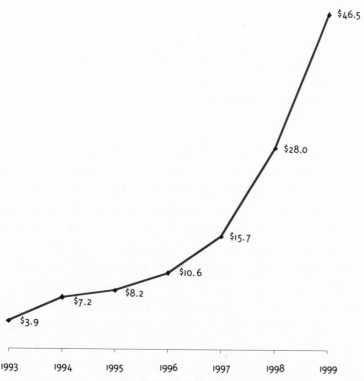

Figure 5-2. U.S. Venture Capital Raised, 1993–1999, in Billions of Dollars
Source: National Venture Capital Association

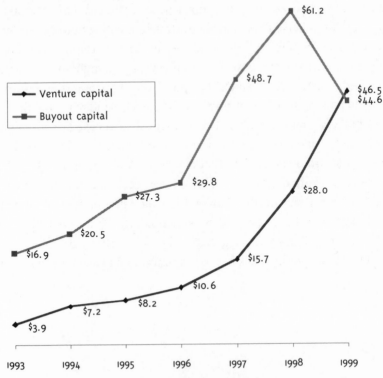

Figure 5-3. U.S. Venture and Buyout Capital Raised, 1993–1999,
in Billions of Dollars
Source: National Venture Capital Association

1999 represents an important watershed event for three reasons. First, it suggests that it took institutional investors—the largest component of investors in venture capital firms—about four years to go from realizing that venture capital offered higher returns to actually shifting their money away from buyout funds and into venture capital. Second, it raises concerns that this shift from buyout funds to venture capital funds could signal the end of the highest rates of return as too much money chases after too few good deals. Third, the crossover passes the baton of "master of the universe" from the buyout funds, such as Kohlberg, Kravis and Roberts, to the venture capital funds, such as Kleiner, Perkins, Caufield and Byers. In fact, this passing of the baton has been signaled by a trend that began in 1999 of buyout fund partners investing their personal funds in Internet startups.

There has been a significant amount of change in the industries in which venture capital firms have been investing. As Figure 5-4 illus-

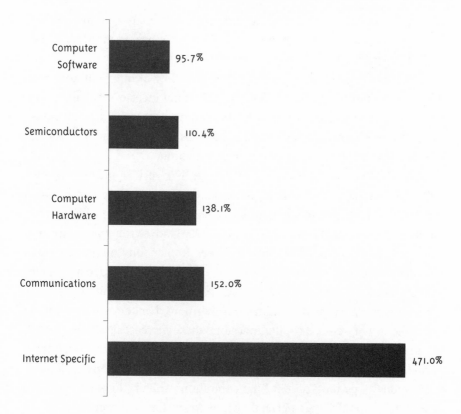

Figure 5-4. Percent Change in Venture Capital Invested by Sector, 1998–1999
Source: National Venture Capital Association

trates, Internet-specific investments grew far faster than other investment categories between 1998 and 1999. As Figure 5-5 (page 103) indicates, Internet-specific investments were also the largest recipients of venture capital in 1999, followed by communications, which received less than half of Internet-specific's share of venture capital.

These results suggest an important insight for investors in Internet venture capital firms, as we suggested earlier: There is a significant time lag between the recognition of high returns in a particular sector and the flow of investment capital into that segment. What this time lag suggests is that the highest returns are made by the "smart money" and that the "dumber money"—of which there is a larger quantity—tends to follow the sectors where the smart money achieved considerable investment returns. The question for investors in Internet venture capital firms is whether it makes sense to precede or to follow the dumber money. As the dot-com consolidation continues, it will become clear that many venture investors look dumber in retrospect than they might have appeared at the time they made their initial investments.

The fundamental reason for the increased flow of into venture capital has been the returns generated for venture capital investors by initial public offerings. As Figure 5-6 (page 104) demonstrates, the amount raised through these IPOs has grown dramatically, as has the spread between the offered price and the subsequent market valuation.

Clearly the post-offer valuation has grown very dramatically since 1997. In that year, for example, the amount of the average pop between initial offering price and post-offer valuation was 416%. This amount climbed to 466% in 1998 and 577% in 1999. Underwriters recognized in 1998 and more so in 1999 that there was tremendous marketing value associated with huge spreads between the initial offering price and a subsequent market value. Simply put, the press enjoys writing about enormous one-day pops in IPO values. Furthermore, bankers enjoy bragging about them. For venture capital investors, these increases are fine as well, because these investors are typically required to hold their shares for a period of about 180 days after the initial public offering. Thus, as long as the one-day pops hold up until the expiration of their "lock-up" periods, the venture capitalists are perfectly happy to see this outcome.

While all this is good news for venture capital firms, the one party that suffers is the IPO company itself. The company suffers because the one-day pops leave a lot of money on the table that does not make it into the coffers of the firm whose shares are being sold to the public. While this lack of corporate cash is not a problem during a period when investors' appetite for Internet shares cannot be sated by the supply, it does become a significant problem when there is a reversal of fortune. More specifically, when investors lose interest in purchasing additional shares of cash-consuming Internet firms, then the relatively low amount of cash that they raised at the offering price shortens their life span considerably.

From the perspective of economic bargaining power, the tremendous wealth that venture capital firms create helps to strengthen their negotiating position when facing both their limited-partner investors and entrepreneurs seeking capital. As Figure 5-7 (page 105) indicates, as of March 1999, the financial returns earned by early/seed venture funds was significantly higher than returns generated by other forms of private equity. Fur-

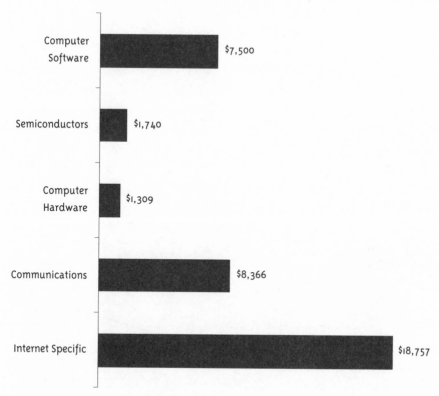

Figure 5-5. Venture Capital Investments by Sector, 1999, in Millions of Dollars
Source: National Venture Capital Association

thermore, these superior returns held up over one-year, three-year, and five-year periods, only falling slightly behind later-stage venture funds over a 10-year period.

These high returns are generated largely by the process of financing early-stage ventures that gives venture capital firms enormous returns from their companies that make it to the IPO stage. Venture capital industry lore has it that a venture firm will review 20,000 business plans a year and invest in only 10 or 20 of the plans. For every 10 firms in which the venture capital firm invests, one will be a huge winner and the balance will be either mediocre performers or total losses. In order to generate the average returns illustrated in Figure 5-7, the value of that one spectacular success must make up for a significant amount of failure.

What the public does not generally realize is that the process of

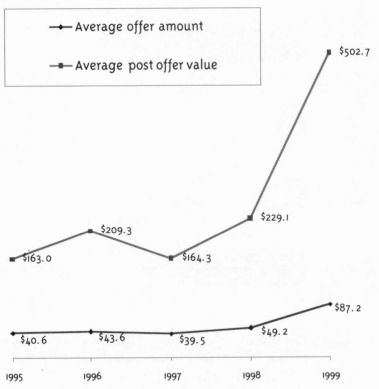

Figure 5-6. Average Venture-Backed Initial Public Offering Amount and Post-Offer Value, 1995–1999, in Millions of Dollars
Source: National Venture Capital Association

financing early-stage ventures tilts the playing field dramatically in favor of spectacular investment returns. When a new venture is financed, it typically raises four rounds of funding. Each round is generally structured as a series of convertible preferred equity. This equity is generally converted into shares of common stock at the initial public offering. Companies typically issue the rounds of preferred equity at increasingly higher prices, which reflect higher valuations for the private company. Figure 5-8 (page 106) illustrates how valuations increased over time during the process of funding an online printing service company called Noosh, which filed for an initial public offering in April 2000 (at the time of the filing, Noosh's total revenues were nil). Noosh withdrew this IPO after the April 2000 dot-com market crash.

Nevertheless, this example illustrates the economic power the venture

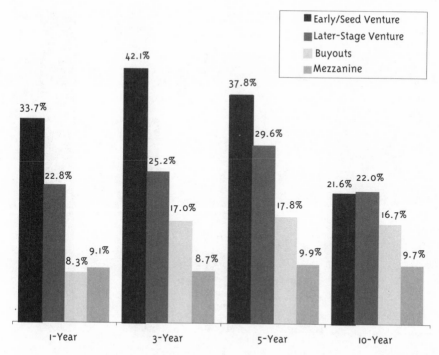

Figure 5-7. Average Investment Returns for Selected Private Equity Fund Types as
of 3/31/99 for 1-, 3-, 5-, and 10-Year Investment Horizons
Source: Venture Economics

capital firms wield. Typically, the venture capital firms invest heavily at the Series A round and maintain their proportionate shares of the company in each subsequent round. Here is a simplified example where a venture capital firm buys one million shares of Noosh in the Series A round and simply holds the shares until the IPO, which we will assume has an initial offering price of $12 per share and doubles on the first day of trading to $24 per share. For purposes of this example, we will assume that end share of Series A, B, C and D equity is converted into one share of common stock—the security that is issued in the initial public offering.

Typical returns for a successful early-stage venture capital firm can be huge if the venture goes public. Based on the simplified example outlined above, let's assume a venture capital firm writes a check for $650,000 for one million shares of Noosh Series A in November 1998. At the Series B funding round in April 1999, the Series A investment increases in value by 423% to $2.75 million. At the Series C funding in November 1999, the

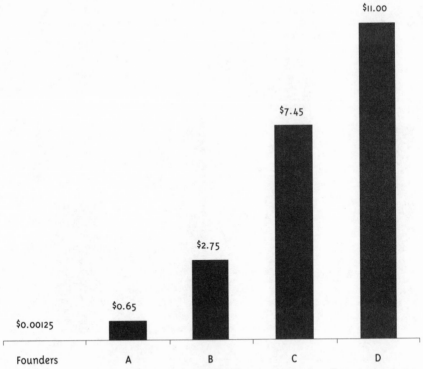

Figure 5-8. Noosh, Inc., Price per Share of Founders' Shares and Four Convertible
Preferred Rounds, 1998–2000

Source: Noosh, Inc., Prospectus, 4/7/00

Series A investment grows by 11,462% to $7.45 million. At the Series D round in January 2000, that Series A investment increases in value by 16,923% to $11 million. The initial offering price of the common equity (assume that it happened in May 2000) represents an 18,462% increase over the value of the Series A investment, producing a total of $12 million, doubling on the first day to $24 a share. This is an increase from the initial Series A investment of 36,923% to $24 million in 18 months. With returns this high, there are undoubtedly some enormous losses out there to lead to an average three-year return on early/seed stage venture funds of a mere 42.1%.

Internet venture capital is an industry with very high levels of economic bargaining power. It is large and growing fast, and it generates enormous investment returns. The primary challenges it faces are the emergence of new entrants (including corporate venture funds and "angel" investors), the uncertain market for IPO, and a potential decline in the quality of the deals financed as a result of the huge influx of money chasing a limited number of outstanding investment opportunities.

Interestingly, a temporary decline in the market for IPOs, as experienced in much of 2000 and potentially 2001, could help venture capitalists deal with the issue of limited investment opportunities. If the IPO market remains weak for a long time, many poorer-quality venture-backed Internet companies will go out of business, causing some of their venture capital backers to leave the venture capital business. This could leave the venture capital field to fewer, more successful venture capital firms able to finance the higher-quality deals with less competition.

Closed-Loop Solution

One of the most important strengths of the leading venture capital firms is their ability to offer their portfolio companies a closed-loop solution. In venture capital, a closed-loop solution means much more than simply writing a check. Besides capital, startup companies rely on venture capital firms to help them through their networks in many different areas, including legal, accounting, building an effective management team, finding customers, negotiating with partners, and working with underwriters to complete their initial public offerings.

While there are not too many opportunities for individual investors to tap into the profits generated by these closed-loop solutions—we have already noted that there are only two publicly traded Internet venture capital firms—the techniques that the leading venture capital firms use are worth understanding. It is interesting to note that many of the leading venture capital firms—as measured in terms of their share of IPO proceeds—are in fact not what we would think of as traditional venture capital firms. More specifically, many of the leading venture capital firms are in fact subsidiaries of investment banks and technology companies.

To get a sense of which firms have achieved the most success in venture investing, it is worth noting that 1999 was the banner year for IPOs, but it was followed by a notable decline in the first quarter of 2000. Indeed, the decline has continued as a result of the April 2000 market crash in technology stocks, which saw a 50% decline in the NASDAQ from its March 2000 high of 5,000. Figure 5-9 highlights the IPO market decline in the first quarter of 2000.

In the context of the declining overall volume, it is useful to understand

which firms were able to grab the largest share of the shareholder value created and which were able to generate the highest investment returns. Table
5-1 summarizes the results of 10 leading venture firms, ranked in descend-
ing order of the number of IPOs conducted during the first quarter of 2000.

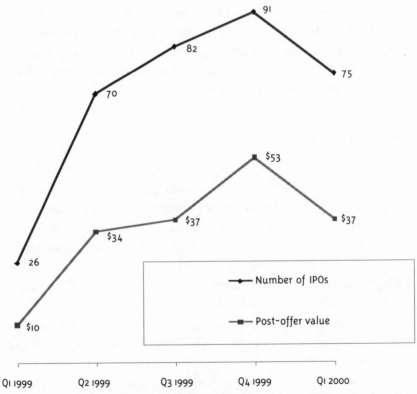

Figure 5-9. Number and Post-Offer Value of Venture-Backed Initial Public
Offerings, in Billions of Dollars (First Quarter 1999 to First Quarter 2000)
Source: National Venture Capital Association

TABLE 5-1. NUMBER OF IPOS, OFFERING AMOUNT, AND POST-OFFER VALUATION, FIRST QUARTER, 2000, IN MILLIONS OF DOLLARS

Firm Name	Number of IPOs	Total Offer Amount	Total Post-Offer Valuation
Hambrecht & Quist Private Equity	9	$1,823.5	$13,021.0
Burr, Egan, Deleage & Co.	6	2,528.1	12,613.0
Bessemer Venture Partners	5	1,919.4	13,562.0

Firm Name	Number of IPOs	Total Offer Amount	Total Post-Offer Valuation
Mayfield Fund	5	1,703.7	12,535.0
Alta Partners	4	669.8	1,601.5
Weiss, Peck & Greer . Venture Partners, L.P	4	872.0	4,599.7
Microsoft	4	651.1	3,383.4
Oak Investment Partners	4	1,289.0	3,579.6
Bowman Capital	4	453.0	2,212.3
Goldman Sachs & Co.	4	474.8	3,570.4

Source: Venture Economics and National Venture Capital Association (NVCA)

Note: Post-offer value is the total market capitalization of the companies going public at the offer price.

An analysis of this table indicates that one way for public investors to gain a share of the returns might be to buy shares in their parent companies. For instance, Hambrecht & Quist Private Equity is a subsidiary of publicly traded Chase H&Q, Microsoft's venture gains are part of publicly traded Microsoft, and Goldman Sachs' gains are part of that firm's total profit. The challenge in this case is that investors in the parent companies cannot participate in the returns that direct investors in the subsidiaries earn. Other banks such as Morgan Stanley Dean Witter and such technology companies as Cisco Systems and Intel have venture capital arms, but the problem for individual investors is their inability to gain direct access to the returns of the venture capital portfolio.

Simply put, the returns of these firms' venture portfolios are dwarfed by the financial performance of the parent companies' other subsidiaries. Investors should understand the relationship between the parent companies and their venture capital subsidiaries. For example, Chase is able to offer a number of important services to its venture-backed companies. Chase provides access to debt financing through its banking subsidiaries, and its Hambrecht & Quist investment banking unit provides access to the IPO market. In addition, its private banking units can offer the newly wealthy executives of the venture-backed companies personal financial planning and other services. Goldman Sachs and Morgan Stanley are able to provide the same advantages, with the added benefit to the average venture-backed company of having a much more highly regarded underwriting capability.

In fact, Chase H&Q is not among the top-10 IPO underwriters; Morgan Stanley and Goldman Sachs occupy the number one and two positions, respectively. According to the *Wall Street Journal*, between January 1, 1999, and March 31, 2000, Morgan Stanley's IPO underwriting business generated $20.23 billion worth of proceeds from 60 deals, while Goldman Sachs's business generated $19.09 billion worth of proceeds from 71 deals (Ewing, 2000). As Figure 5-10 illustrates, the stock market performance of these offerings has been unimpressive, at least in the first quarter of 2000.

Investors should also realize that corporate investors such as Microsoft and Cisco Systems each offer very different parts of a closed-loop solution for startup companies. Cisco Systems—as we noted in Chapter 4—has made acquisitions an important part of its growth strategy. Often Cisco Systems acquires privately held companies in which it may have been a venture investor. Such venture investing provides the new companies with ancillary benefits, including access to customers, management talent, manufacturing, distribution, administrative infrastructure, and the capital markets.

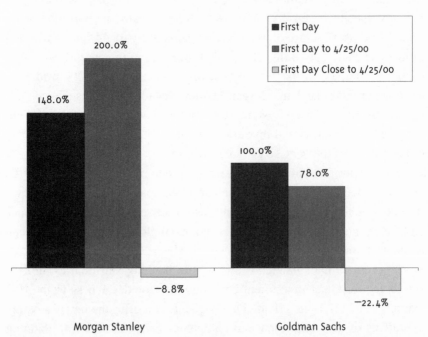

Figure 5-10. Morgan Stanley and Goldman Sachs, Percentage Change in IPO Prices: First-Day Gain, from Offer Price to 4/25/00 and from First Day Close to 4/25/00
Source: Wall Street Journal

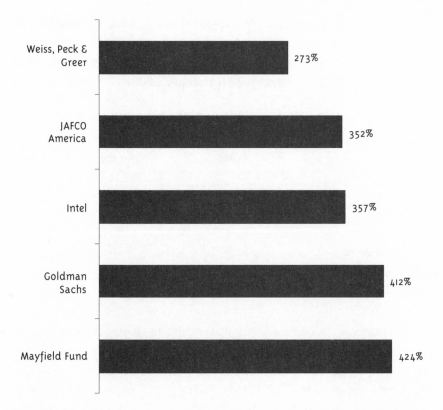

Figure 5-11. Selected Venture Capital Firm Post-IPO Returns of First Quarter 2000 IPOs, as of March 31, 2000

Source: Venture Economics and National Venture Capital Association

Note: Total return based on $1,000 invested in each 2000 Q1 IPO at the offer price.

Often the corporate investors in new ventures can generate significant returns on their portfolios. As Figure 5-11 indicates, these returns can be extremely high and can be generated in very short periods of time.

The year 2000's stock price declines by publicly traded venture capital firms (see Table 5-2) highlight weaknesses in their ability to provide closed-loop solutions. During their heydays CMGI and ICGE touted the benefits of providing management expertise and an internal market for each firm's products or services. According to the *Red Herring,* the hollowness of these benefits was revealed by their late 2000 restructuring activities. In November 2000, ICGE laid off 50 employees, or 35% of its staff, and announced plans to take a $25 million to $30 million charge in the fourth quarter of 2000 as a result. CMGI merged two of its partner companies, MyWay.com and Zip2, resulting in an undisclosed number of layoffs in early October 2000. CMGI also said it planned to stop funding entertainment partner company iCast and advertising-supported ISP 1stup.com by the end of January 2001 (Landry, 2000).

TABLE 5-2. CMGI AND INTERNET CAPITAL REVENUE MOMENTUM, 2000

$ in Millions	CMGI	Internet Capital Group
Stock percent change	−96%	−98%
12 months' revenue	$1,140	$24
Revenue growth	184%	47%
Market capitalization*	$1,840	$1,410

Source: MSN MoneyCentral

*As of January 17, 2001.

The publicly traded venture firms were able to go public because retail investors saw them as the closest that they would be able to get to the huge investment returns that venture capitalists were making on dot-com IPOs. The more companies in ICGE's and CMGI's portfolios (and the more difficult they were to value), the easier it was to create a momentum-induced buying panic for their shares. The dot-com crash in April made it clear that the CEOs of these companies were not geniuses and that they were simply presiding over a hugely unprofitable collection of businesses, most of which had no chance of ever reaching the stage where they could go public. Simply put, the touted management skills and partnership benefits of being part of the ICGE's and CMGI's portfolio were empty promises. In response, the firms undertook restructurings.

CMGI's stock suffered through most of 2000; however, it took until August for CMGI to take action publicly. In September 2000, CMGI announced that it would reduce its portfolio of 17 companies to between five and 10 by July 2001. This portfolio reduction, while consistent with corporate restructuring practices, completely undermined the logic of management skills and portfolio synergies on which CMGI's portfolio was originally built. Meanwhile, @Ventures, CMGI's private venture capital unit, closed several of its portfolio companies, including Motherna-ture.com (which had gone public), Furniture.com, and Productopia. With the unwinding of the logic that supposedly made the value of the whole of

CMGI greater than the sum of its parts, it remains unclear how CMGI will fare going forward.

ICGE is undergoing a restructuring as well, albeit a less clearly articulated one. ICGE has categorized its companies into "developed" and "emerging." ICGE intends to prioritize its resources in favor of the "developed" companies, but as of November 2000 it had not specified how many are in that category. The companies in ICGE's "developed" category plan to be profitable within the next 18 months and to show pretax profits of more than $100 million in two to five years, and have strong management teams. While these criteria are typical of a company that could have gone public prior to the April 2000 crash, by November 2000, ICGE's CEO suggested that even he did not expect the IPO market to return to a level that would support IPOs for B2B companies in 2001—casting significant doubt on ICGE's prospects.

Assessing the strategies of publicly traded Internet venture capital firms remains difficult in practice. The stated efforts to offer portfolio companies with a closed-loop solution sounded plausible in 1999. However, by 2000, the collapse of the IPO market made it crystal clear that the management skill and network synergies of these firms' portfolios were of ephemeral value. The stock prices of these publicly traded Internet venture capital firms had collapsed by the end of 2000. This collapse forced the companies to close portfolio companies to conserve cash—while attempting to make it simpler for investors to evaluate the quality of their strategies.

It is almost impossible to discern how private VC partnerships are dealing with the loss in value of their dot-com portfolio companies. Probably, their restructuring efforts parallel those of their publicly traded brethren.

Management Integrity and Adaptability

Management integrity and adaptability are just as important in Internet venture capital as they are in other industries. Since there are so few publicly traded Internet venture capital firms, it is much more difficult to judge these attributes. One of the most significant integrity issues with venture capital firms is a lack of respect for intellectual property. For example, there is anecdotal evidence that some venture firms refuse to sign nondisclosure agreements (NDA) before meeting with an entrepreneur.

More than once, an entrepreneur has pitched a business plan to a venture capital firm without receiving a signed NDA in return. Subsequently, that entrepreneur has found that the non-NDA-signing venture capitalist has invested in a company whose business plan is remarkably similar to the one pitched by the entrepreneur. While the absence of a signed NDA makes this ethical violation difficult to prosecute, it provides suggestive evidence of the kinds of integrity breaches that can occur between a venture capital firm and the entrepreneur seeking its assistance. It should not be surprising that a venture capitalist willing to breach the trust of an entrepreneur may be equally willing to breach the trust of an investor as well.

Financial statements provide clues about management integrity. As Table 5-3 indicates, the financial statements of CMGI raise some questions regarding its management integrity. CMGI is subject to a number of lawsuits, including ones that allege a breach of contract with an executive and a business partner. These lawsuits are significant in that they suggest that CMGI management might have a somewhat high-handed attitude toward its staff and partners—and possibly its investors.

Internet Capital Group may have some imprecise revenue measurements but has not been party to any legal action. Internet Capital Group's VerticalNet subsidiary engages in barter transactions, which are based on estimated fair values of advertising services. These estimates could be subject to some variations that could lead VerticalNet's revenues to be over- or understated. Internet Capital Group's absence of specific litigation could be a sign of relatively high levels of management integrity.

TABLE 5-3. REVENUE ACCOUNTING POLICY AND SHAREHOLDER LAWSUITS FOR CMGI AND INTERNET CAPITAL

	CMGI	Internet Capital
Revenue Accounting	CMGI advertising revenue is recognized in the period the advertising impressions are delivered, provided the collection of the resulting receivable is probable. Prior to August 1, 1998, revenue from the sales of product licenses to customers were generally recognized when the product was shipped,	All of the ICGE's revenue during 1997 and 1998 was attributable to VerticalNet. VerticalNet's revenue is derived from advertising contracts which include the initial development of storefronts. A storefront is a Web page posted on one of VerticalNet's trade communities that provides

	CMGI	Internet Capital
	provided no significant obligations remained and collectibility was probable. Revenue from usage-based subscriptions is recognized monthly based on actual usage. Service and support revenue includes software maintenance and other professional service revenues, primarily from consulting, implementation, and training. Revenue from software maintenance is deferred and recognized ratably over the term of each maintenance agreement, typically 12 months. Revenue from professional services is recognized as the services are performed, collectibility is probable and such revenues are contractually nonrefundable.	information on an advertiser's products, links a visitor to the advertiser's Web site, and generates sales inquiries from interested visitors. Advertising revenue is recognized ratably over the period of the advertising contract. Barter transactions are recorded at the lower of estimated fair value of goods or services received or the estimated fair value of the advertisements given. Barter revenue is recognized when the VerticalNet advertising impressions are delivered to the customer, and advertising expense is recorded when the customer advertising impressions are received from the customer.
Legal Proceedings	Neil Braun, the former president of iCast Corporation, a majority-owned subsidiary of CMGI, filed a complaint against CMGI alleging certain claims arising out of the termination of Mr. Braun's employment with iCast. Mr. Braun seeks $50 million in damages. International Merchandising Corporation filed a complaint against Signatures Network, Inc., a subsidiary of iCast and CMGI. The complaint asserts breach of contract and seeks damages of $15 million plus stock options.	None

Source: Company reports.

Publicly traded Internet venture capital firms have not demonstrated the ability to adapt very effectively to change, as measured by the variance of the percentage change in their quarterly earnings. Simply put, publicly traded Internet venture capital firms have demonstrated great volatility in their earnings. In part, this is a function of the extremely uneven timing of the realization of investment gains.

Another factor is the inability of these firms to adapt quickly enough to the changes in the Internet industry. In effect, Internet venture capital firms must make bets on companies before it is clear whether the companies' business strategy or industry will generate significant revenues and profits. As a result, it is often the case that a bet will prove to be incorrect. By the time it becomes clear that the bet was wrong, it is too late to undo the investment. Fortunately, as noted earlier, for every 10 investments made by the average venture capital firm, one investment offers such high returns that it makes up for the less-than-stellar performance of the other nine.

As noted earlier, the earnings changes for CMGI and Internet Capital have been very volatile. As Figure 5-12 indicates, however, CMGI's performance has been significantly more volatile than that of Internet Capital. Nevertheless, given the inherent difficulty of producing predicable earn-

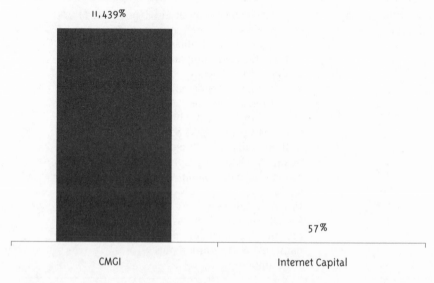

Figure 5-12. Variance in Annual Earnings Percent Change for Selected Internet Infrastructure Firms, 1995–1999

Source: MSN MoneyCentral, Peter S. Cohan & Associates Analysis

ings for a publicly traded venture capital firm, investors should not read too much significance into these results.

The fundamental reality is that management adaptability in Internet venture capital can only be measured over much longer periods of time—such as the three- to five-year cycles that are typical of the elapsed time between when a venture capital investment is made and when the company is taken public. It is very difficult for momentum investors to cope with these longer time frames for assessing management adaptability. As a result, the short-term stock market performance of Internet venture capital firms is likely to continue to be difficult to relate to their business fundamentals. A significant reason for this difficulty is that the financial statements of publicly traded venture capital firms are not very useful tools for assessing the worth of the investment portfolios.

Management integrity and adaptability are critical factors in differentiating winning Internet venture capital firms from their peers. Our analysis of CMGI and Internet Capital suggests that CMGI might be a bit weaker in this category than Internet Capital. Unfortunately, the tools available to investors for assessing publicly traded Internet venture capital firms leave much to be desired. It would be useful if these firms could report their investment returns in the way we reviewed earlier in this chapter for the funds managed by Intel and Goldman Sachs (Figure 5-11). Of course, what would be most valuable for investors would be for management to detail its views of the six IID indicators we are reviewing in this chapter for each portfolio company. Regrettably, the odds of such reporting actually being required by the SEC are extremely low.

Brand Family

Internet venture capital firms are typically judged by the general public on the basis of the performance of their best-known investments. For example, Kleiner Perkins is known for its investment in Netscape, Sequoia is known for investing in Cisco and Yahoo. The publicly traded venture capital firms have a mix of successful and less successful companies in their portfolios. As Table 5-4 illustrates, CMGI's publicly traded subsidiaries have turned in a mixed performance. Navisite, a Web-site hosting-services provider, increased 400% between its IPO and its March 2000 high but plummeted from this $150 level to $3 in November 2000, whereas Mothernature.com, an online distributor of vitamins and provider of health formation, saw its stock price collapse by 76% since its IPO and subsequently liquidated itself in November 2000. Internet Capital's most prominent subsidiary, VerticalNet, saw its stock price double since its April 1999 IPO. While VerticalNet has remained significantly above water, its stock price has declined 200% from its all-time high.

TABLE 5-4. STOCK MARKET PERFORMANCE OF CMGI AND INTERNET CAPITAL SELECTED PORTFOLIO COMPANIES, PERCENT CHANGE IN PRICE ON FIRST DAY IPO CLOSE TO 11/28/00

Company	Portfolio Company (Stock Price % Change)
CMGI	Navisite (−77%)
	Engage Technologies (−90%)
	Critical Path (−74%)
	Mothernature.com (bankrupt)
Internet Capital	VerticalNet (−59%)
	Breakaway Solutions (−85%)
	Onvia (−97%)

Source: Company Financial Statements, MSN MoneyCentral

Judging by the performance of the companies in their portfolios, CMGI and Internet Capital have performed disastrously. As a consequence, these firms' brand families are likely to inspire investors to flee these stocks. The value destroyed in CMGI's portfolio companies has stripped David Wetherell, CMGI's chairman, of the star power he enjoyed in the media during much of 1999. Wetherell's star power has been more starkly dimmed by the 85% decline in CMGI's stock between the end of 1999 and the end of 2000.

ICGE's star has also been dimmed. Internet Capital Group's ability to attract many luminaries to its advisory board helped the firm's media perception, as did *The Industry Standard*'s April 2000 selection of Ken Fox as the most effective B2B executive (Evans, 2000). Nevertheless, the 91% decline in the stock price of Internet Capital Group and the collapse of Onvia stock since its IPO reflect diminished enthusiasm for B2B in general and may make it difficult for Internet Capital Group's stock to recover.

The privately held venture capital firms such as Kleiner Perkins,

Sequoia, and Draper Fisher Jurvetson have created more compelling brand families. Unfortunately, investors in public equities cannot tap into the returns of these privately held funds. So the publicly traded Internet venture capital firms currently have terrible brand families, while some of the privately held Internet venture capital firms enjoy great reputations, which investors in public equities cannot tap into. During times (such as the latter half of 2000) when portfolio companies collapse, the reputational benefits to venture capitalists of losing money in private become clearer.

Financial Effectiveness

It is useful for investors to know how Internet venture capital firms are using their financial resources. While investors care most about the stock market performance of a firm's portfolio companies, useful insights can be gained from examining these companies' sales productivity and overhead ratios, since these numbers provide insights into the performance of the portfolio companies whose financial statements are consolidated into the parent companies.

Both firms have similar levels of sales productivity, but Internet Capital Group is spending a large proportion of its cash flow on developing the company. As Table 5-5 suggests, CMGI and Internet Capital have virtually identical levels of sales per employee. In the case of CMGI, these sales represent the operating revenues of its portfolio companies to the extent that their sales flow through to CMGI's financial statements and to the marketing-services business that is a major part of CMGI's financials. For Internet Capital Group, the sales figure represents only the operating revenues of its portfolio companies.

Both CMGI and Internet Capital have high ratios of SG&A to revenues. In the case of CMGI, this is due to the operating expenses of its main operating business, providing marketing services. For Internet Capital Group, this high ratio reflects the huge amount that its portfolio companies are spending on sales, marketing, and advertising in order to establish a large market share. While neither of these firms is demonstrating frugality in the use of financial resources, the level at which they are spending money is not unusual for the industries in which they compete.

TABLE 5-5. CMGI AND INTERNET CAPITAL GROUP SALES PER EMPLOYEE AND SG&A/SALES RATIOS

Company	Sales/Employee	SG&A/Revenues
CMGI	$236,198	76%
Internet Capital	$235,714	296%

Source: MSN MoneyCentral

Relative Market Valuation

While the foregoing factors are all important for the evaluation of Internet venture capital stocks, they are all precursors to the most fundamental measure—the relative value of the firms from the perspective of the securities market. The fundamental question facing investors is whether a stock that is trading at a relatively low ratio of sales or book value is more or less likely to go up in value. Simply put, investors need a way to determine whether the market is marking down a particular stock because it is less likely to go up in value or more likely to go up in value.

Having examined the five other gauges on the Internet Investment Dashboard, we are now in a better position to answer this question. As Table 5-6 indicates, the market seems to be sending mixed signals regarding the relative values of CMGI and Internet Capital Group.

TABLE 5-6. STOCK PRICE AND REVENUE GROWTH, PERCENT CHANGE, 2000, AND VALUATION RATIOS FOR CMGI AND INTERNET CAPITAL, JANUARY 2001

Company	Stock % Change	Revenue Growth	Price/ Sales	(Price/Sales)/ Rev. Growth	Price/ Book	Market Cap./ Portfolio Company
CMGI	−96%	184%	1.2	0.7	0.27	$322 million
ICGE	−98%	49%	46.1	94.1	0.40	$24 million

Source: MSN MoneyCentral

On the basis of price/sales and price/book, Internet Capital Group appears to be highly valued relative to CMGI, despite ICGE's 98% drop in price since the end of 1999. From the perspective of the stock market value of the average holding in the portfolio, however, CMGI appears to be more highly valued by the market. Given the distortions associated with the revenue (such as the inclusion of CMGI's substantial marketing-service revenues) and book value figures (such as the understatement of the intangible assets), the market capitalization/portfolio company measure may well be the most pertinent one.

The Internet Investment Dashboard analysis suggests that publicly traded Internet venture capital firms should be approached with caution. In Figure 5-13, two of the indicators on the dashboard are clear, but most are dotted. These indicators suggest that investors may wish to avoid CMGI and Internet Capital. In the future, there may be more ways for investors to gain access to this sector; however, it is likely to be several years before the conditions for such investment are ripe.

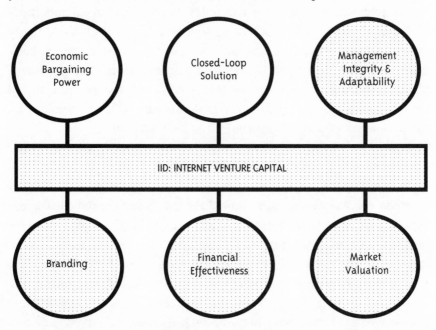

Figure 5-13. Internet Investment Dashboard Assessment of Internet Venture Capital Segment

Key: Clear = positive performance and prospects; dotted = mixed performance and prospects.

PRINCIPLES FOR INVESTING IN INTERNET VENTURE CAPITAL STOCKS

Needless to say, the analysis we have conducted here is subject to change. Undoubtedly some of the weaker firms could generate positive surprises, and new competitors could emerge to tilt the playing field in a totally different direction. As a result, investors in Internet venture capital stocks need some principles to guide them through the analytical process as investment conditions evolve.

Internet Venture Capital Investment Principle 1. Monitor developments in new technologies that can represent new opportunities for venture investing. Investors must recognize that excellent Internet investment opportunities change very rapidly. Specific sectors go in and out of fashion very quickly. Investors must look beyond the fads to understand which businesses are likely to prosper over the long term and which ones are vulnerable. In addition, investors must assess how well the publicly traded venture capital firms succeed in capturing the value of their foresight, by measuring the relative performance of their investments.

Internet Venture Capital Investment Principle 2. Analyze the evolution of the upstarts' strategies to assess which will be able to offer more value to portfolio companies. Investors should recognize that incumbent venture capital firms enjoy tremendous economic bargaining power with entrepreneurs. As new models for matching investors with entrepreneurs become more prominent, such as the online capital networks for firms like Round1, there could well be a shift in which venture firms are able to finance the best deals. Investors in publicly traded Internet venture capital firms should monitor these shifts in value propositions to entrepreneurs and invest in the Internet venture capital firms that offer the biggest incremental improvements.

Internet Venture Capital Investment Principle 3. Avoid investing in firms where there are questions about management integrity and adaptability. Our analysis indicated questions about management integrity and adaptability in a few firms. This analysis is most useful for investors as a sell signal. If a firm with a track record of avoiding shareholder lawsuits and generating consistent profit growth suddenly deviates from this path, investors should consider selling the stock. In many cases, such deviations from the path of integrity and adaptability could be a signal that management is no longer up to the job. When these deviations occur, it is likely that management will try to gloss over the problems. Investors may not want to stay along for the ride.

Internet Venture Capital Investment Principle 4. Use compelling brand families as a signal of value in conjunction with other indicators. Some of the firms we studied had less than stellar investment track records—at least based on the most recent performance of their portfolio companies in the stock market. Investors should monitor the performance of the portfolio firms in the stock market and recognize that the market will gravitate toward the firms with the best overall portfolio performance. The challenge facing investors is that the accounting for Internet venture capital firms does not make it as easy to look at them as it is to look at mutual funds.

Internet Venture Capital Investment Principle 5. Recognize that the accounting for publicly traded Internet venture capital firms does not lend itself to incisive analysis of financial effectiveness. The accounting techniques used for publicly traded Internet venture capital firms are difficult to decipher. Investors in these firms are most interested in understanding the investment prowess of their managers. Unfortunately, their financial statements make this performance difficult to assess. Investors should focus on valuing the stakes that the Internet venture capital firms hold in their portfolio companies. While these evaluations are relatively easy to accomplish for publicly traded holdings, they fall short for privately held ones unless there are true comparable companies available.

Internet Venture Capital Investment Principle 6. Monitor relative valuations on the basis of the average valuation per portfolio company. Valuing Internet venture capital firms on the basis of price/sales or price/book does not make that much sense given the distortions in the denominators of these ratios. As a result, the most practical valuation measure may well be the relative value of the average portfolio company. To the extent that these measures can be compared to industry averages, investors may find the relative valuation measures useful.

CONCLUSION

Despite the drop in returns in 2000, Internet venture capital is an attractive area for long-term investment; however, the publicly traded Internet venture capital firms may not be the best ways to take advantage of the opportunities. While the industry has economic leverage and there are firms with good strategies, variations in the quality of management and of brand families make it important for investors to assess the six indicators to identify the individual firms within the industry that are likely to prevail. Valuation matters for this segment, and depending on the investment time horizon, investors may want to watch closely for opportunities to purchase shares in segment market leaders during periods of market decline in which excellent companies are punished along with poor ones. Such well-timed investments could yield attractive returns.

Chapter 6

PILEUP ON THE INFORMATION SUPERHIGHWAY: WEB CONSULTING STOCK PERFORMANCE DRIVERS

W EB CONSULTING IS an industry that depends heavily for the extent of investment returns on the level of demand for its services. When e-commerce startups and land-based firms are eager to build e-businesses, the Web consulting firms' stock prices as well as their revenues (and in some cases profits) grow very rapidly. If the demand for help building Web sites drops off, the stock prices of Web consulting firms can fall fast along with the growth rate in their revenues. Furthermore, not all Web consulting firms do equally well in terms of their financial performance or their stock price appreciation. This chapter will help investors distinguish the winners from the losers.

The April 2000 crash caused a drop in demand for Web consulting services. Dot-coms no longer had the cash to pay for services. Land-based firms lost their sense of urgency for deploying Web cousulting services as well. By November 2000, Web consultants were retrenching to conserve cash. One such firm, MarchFirst, laid off 1,100 employees to save $100 million. However its ability to meet payroll was in doubt.

All Web consulting firms saw their stock prices crumble in the metal-screeching pileup that followed the April 2000 Internet stock crash. It took about three months from the crash for its effects to work their way through to the Web consultants—many of whom were turning away business from venture-backed dot-coms. Sapient, as a result of its dependence on land-based clients that could continue to pay their bills after April, saw its stock price rise 118%, from $22 in September 1999 to $48 in September 2000. In September 2000, Sapient's success landed its co-CEOs on *Fortune*'s list of the 40 wealthiest individuals under 40 years of age (with net worths of about $1.4 billion per co-CEO). By December 2000, Sapient's stock had tumbled 75%, taking its co-CEOs' net worth to a "mere" $350 million.

By contrast, consider the collapse of Viant, an e-consultant firm based near Sapient in Boston, whose stock price dropped almost 70%, from $25 in September 1999 to $4 in December 2000. Considering that Viant's stock had been as high as $63, in December 1999, some unfortunate investors in Viant have suffered through a 94% loss in value. One key difference between Sapient and its neighbor Viant was the level of dependence on cash-poor dot-coms for revenues. In August 2000, Viant announced that it would lose money instead of earning a profit. Viant's CEO attributed the loss to dot-com clients that did not get expected funding and to Fortune 2000 companies that were slowing down their e-business initiatives in light of the collapse in funding for the dot-coms. Though Sapient is less dependent on dot-coms than Viant, there is probably more to Viant's earnings shortfall than what its CEO claimed. As we will see in this chapter, a big difference is the quality of management.

Web consulting means helping managers figure out what to do about the Web and then building the systems to realize that vision. Web consulting is a business that has grown rapidly as businesses have gained interest in Webifying their strategy and operations. It is important for investors to recognize that the barriers to entry into this business are relatively low, and therefore it is likely that there will continue to be many industry participants. In addition to the new entrants, well-established consulting firms are adding Web-consulting capabilities. Distinguishing between the firms that are likely to sustain profitable revenue growth from their peers is the most important analytical challenge facing the investor who aspires to generate attractive returns investing in the Web consulting sector.

WEB CONSULTING STOCK PERFORMANCE

Web consultants differ in significant ways from other kinds of management consultants, such as strategy consultants or systems integrators. For example, strategy consulting firms such as Bain, McKinsey and Boston

Consulting Group are known for helping CEOs figure out what their corporate goals are, designing strategies to achieve those goals, and making the organizational changes required to implement these strategies. Sometimes these changes translate into figuring out whom to fire or identifying specific companies to acquire.

Traditional systems integrators like Electronic Data Systems (EDS) or Computer Sciences Corporation (CSC) tend to specialize in large systems-development projects such as installing a corporate-wide accounting system. Traditional systems integrators also get involved in outsourcing a company's information technology department. In effect, the systems integrator acquires the client's IT department, reorganizes it, and charges the time and activity performed by the department back to the company.

Web consultants combine elements of strategy consulting and systems integration and add some unique elements. Web consultants need to be able to work with CEOs to help them rethink how the Web will change their competitive strategy, organization, and operations. Web consultants must also have the systems skills required to integrate a client's back-office systems (e.g., accounting, inventory control, supply chain, and administrative) with the Web site. But Web consultants also need to have skills in understanding how customers will use a Web site and designing the site so that it will be appealing to customers.

Because Web consulting requires so many diverse skills—business strategy, user interface design, Web tools, and systems integration—it is important for Web consultants to create a culture that can attract people with different skills and get them to work together effectively. We noted in Chapter 3 that the stock market performance of a cross section of Web consulting firms was up 322% in 1999. This increase compared very favorably with the NASDAQ's 86% increase in 1999, and it was slightly less than the 339% average stock price increase of the companies in our nine-segment Internet stock index in 1999. Figure 6-1 summarizes the stock market performance of the seven Web consulting companies we analyzed.

The firms that did the best in the stock market in 1999 were among the most recent entrants to the Web consulting business. Firms such as DiamondCluster International and Proxicom, whose sole focus has been Web consulting, tended to do better in the stock market than firms that recently changed their organization to compete in the Web consulting business—such as Sapient or Cambridge Technology Partners. While

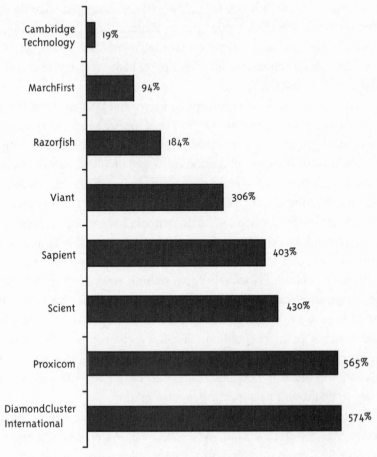

Figure 6-1. Selected Web Consulting Firms' Percent Change in Stock Price,
12/31/98—12/31/99
Source: MSN MoneyCentral

the stock market performance of the Web consulting firms was strong in 1999, as Figure 6-2 indicates, their market value deteriorated significantly in 2000.

The destruction of Web consultants' stock market value was especially hard on those firms that were dependent on dot-coms for their revenues. For example, as we will see later in this chapter, Scient is heavily dependent on consumer-oriented e-commerce startups for its revenues. In April 2000, the crash in Internet stocks turned off the funding spigot for such companies. Prior to April 2000, the dot-coms were involved in a land-grab mentality that forced them to spend money on Web consultants such as Scient in order

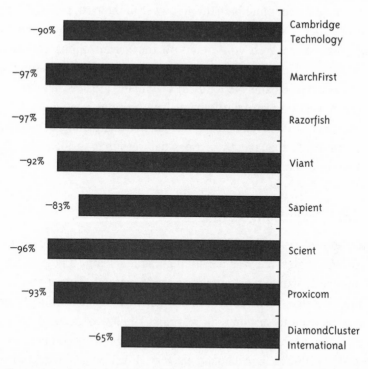

Figure 6-2. Selected Web Consulting Firms' Percent Change in Stock Price,
12/31/99–12/29/00
Source: MSN MoneyCentral

to build their Web sites very quickly and thus get a jump on their competitors. After April, however, the dot-coms found themselves needing to preserve cash. This change in attitude toward cash flow was highly likely to cut into the revenue streams of Scient and its peers. Investors recognized this and sliced huge chunks of value from the consultants' stocks.

INTERNET INVESTMENT DASHBOARD ANALYSIS OF WEB CONSULTING

While the foregoing discussion of the relative stock price performance of the Web consulting segment is suggestive, it does not offer a fully satisfactory explanation. To obtain such an explanation, we apply the Internet Investment Dashboard (IID) analysis to the Web consulting segment.

As we will see, the IID analysis suggests that Web consulting may generate attractive investor returns in the future. While the Web consulting industry enjoyed economic bargaining power in the past, as illustrated by its attractive margins, the postcrash environment had tilted economic bargaining power from Web consultants to the organizations conusuming their services. Selected participants offer customers a closed-loop solution, and these participants are likely to sustain the greatest share of industry profits. Several industry participants have demonstrated their ability to adapt effectively to change and have created a compelling brand family. Many participants manage their financial resources effectively, and a handful are priced relatively cheaply.

Economic Bargaining Power

The Web consulting market has moderate economic bargaining power. As Figure 6-3 illustrates, the Web consulting market is large and growing rapidly. Some participants have attractive profit margins. However, the Web consulting industry suffers from two elements that make it difficult for the average participant to sustain high margins: The low barriers to entry into Web consulting guarantee that there are likely to be many competitors; and the post–April 2000 drop in demand has forced Web consultants to compete

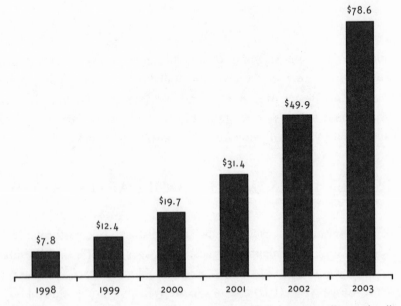

Figure 6-3. Internet Professional Service Market, 1998–2003, in Billions of Dollars
Source: International Data Corp.

more vigorously for fewer new business opportunities. This large number of competitors puts a cap on the profit margins that firms are able to earn.

While there is clearly competition among firms for customers, the more insidious competition takes place for hiring the best Web consultants. There are at least five major sources of competition for the best Web consultants:

- Startup Web consulting firms
- Client/server consulting firms trying to grab a piece of the Web consulting business
- Traditional systems-integration firms trying to revitalize their moribund revenue growth through a shot of dot-com energy
- The dot com startups that dangle attractive stock options and salary packages for top talent
- Bricks-and-mortar companies trying to transform themselves into click-and-mortar operations

The layoffs by Web consulting firms, such as MarchFirst, are likely to dampen the intensity of this competition—enabling land-based firms to take their Web development in-house.

Despite the many competitors, the most successful Web consulting firms have figured out how to earn attractive profit margins. The key to success is the ability of leading firms to offer a unique element of value to customers that enables them to generate fairly steady and profitable growth. One example is Sapient, which provides customers with a fixed-time, fixed-price service that includes all the capabilities we discussed earlier. Sapient's ability to meet its commitments puts it in competition with a narrower group of firms with a similarly strong track record. The limited number of firms that can meet such demanding performance standards enables firms like Sapient to choose clients that are leaders in their industries. The effect of working with leading companies is that firms like Sapient are able to sustain their advantages.

While there are wide variations in the profit margins earned by industry participants, the average Web consulting participant earns negative net margins of −6%. As Table 6-1 indicates, the mean gross margin of seven Web consulting firms is 53%. This relatively high gross margin suggests that the average participant in the industry is able to set a sufficiently high price to offset its costs.

Nevertheless, the negative net margin for the average participant masks a bifurcation in the returns earned by industry participants. There are the money-losers and the money-makers. The money-losers include MarchFirst, Razorfish, and Cambridge Technology Partners. The money-makers include DiamondCluster International, Sapient, Scient, and Viant, with Proxicom at breakeven. While there are many factors accounting for the different profit performance of these two groups, the most significant factor appears to be the different levels of financial controls in the companies. Simply put, DiamondCluster and Sapient are run by managers who keep a very tight rein on costs and revenues, while the money-losing firms are less disciplined.

TABLE 6-1. MARGINS FOR EIGHT WEB CONSULTING FIRMS (DECEMBER 2000)

Company	Stock % Change	Gross Margin	Operating Margin	Net Margin	2-Year Trend
DiamondCluster Int'l	−65%	43%	21%	12%	0
Proxicom	−93	53	5	0	−
Scient	−96	86	2	2	+
Sapient	−83	54	19	11	+
Viant	−92	60	13	12	+
Razorfish	−97	53	5	−2	+
MarchFirst	−97	50	−80	−82	——
Cambridge Tech.	−90	21	−3	−6	−
MEAN	**−89%**	**53%**	**−2%**	**−6%**	**0**

Source: MSN MoneyCentral

Note: Two-year trends represent the extent to which net margins have changed in the most recent two years. + indicates a small improvement, 0 indicates no change, − indicates small deterioration, —— indicates large deterioration.

Table 6-1 also indicates that there is no strong correlation between profit margins and stock price performance. While DiamondCluster International, one of the firms with the highest profit margins, was the best performer in the stock market in 1999, Scient, the second-best performer in the stock market, was among the least profitable of its peer group. Furthermore, the three worst performers in the stock market—Razorfish, MarchFirst and Cambridge Technology—were unprofitable. Simply put, there does not appear to be much of a correlation between the 1999 stock market performance of these eight Web consulting firms and their profit margins.

The Web consulting segment has moderate economic bargaining power. The segment is large and growing at a 59% annual rate. Because of the relatively low barriers to entry into the Web consulting business, there are many competitors, putting a limit on prices and profitability. In addition, there are several strategic groups of competitors vying for a piece of this large and very rapidly growing market. This intense competition makes it unsurprising that the average participant earns net margins of −6%. As we noted earlier, the shift in the locus of bargaining power to buyers of Web consulting services is likely to push net margins even more in the near term. It is likely that the least profitable firms will either close or be acquired—leaving an attractive market for industry leaders should market demand tighten in the future.

Closed-Loop Solution

While it is clear that the Web consulting market is moderately attractive, the next question to analyze is which industry participants offer customers the most attractive value proposition. The concept of the closed-loop solution is particularly important to buyers of Web consulting services because the typical company must make fundamental changes in order to succeed with its e-commerce strategy. Furthermore, companies must conceive and implement these changes very rapidly if they are to maintain their competitive positions.

As a consequence, business managers cannot afford to hire separate firms to rethink their business strategy, their organization, and their business processes. Nor can they afford to hire a new firm to build their Web site and link that site to their back-office systems. To hire separate firms for each of these activities would take too long, in part because of the handoffs between firms, and cost too much money. Even if a company hired separate firms for each of these activities, the quality of the final outcome would probably be weak as a result of the different visions and styles of the various providers.

Vendors that satisfy all these needs under one umbrella are likely to offer customers the highest return on their investment in Web consulting. Offering this high return depends on the firms' relative ability to knit all the services of the Web consulting fabric into a relatively seamless whole—to offer a closed-loop solution. The challenge of the Internet investment analyst is to devise ways to quantify this concept, or at least to gather objective data to assess its existence.

One way to look at the relative effectiveness of the strategies of firms in the Web consulting market is to analyze the firms' relative size, revenue growth, and market capitalization. This analysis can help us gain insights into which firms are generating the greatest product market momentum and the extent to which that momentum has translated into greater stock price appreciation.

The destruction in stock market capitalization during 2000 was extraordinary. As Table 6-2 indicates, the firms that best withstood the decline were not necessarily the fastest-growing ones.

TABLE 6-2. REVENUE MOMENTUM OF EIGHT WEB CONSULTING FIRMS, JANUARY 12, 2001, IN MILLIONS OF DOLLARS

Company	Stock % Change	12 Months Revenue	Revenue Growth	Market Capitalization
DiamondCluster Int'l	−65%	$192	99%	$893
Proxicom	−93	179	181	404
Scient	−96	302	309	230
Sapient	−83	446	88	2,047
Viant	−92	126	172	177
Razorfish	−97	342	368	172
MarchFirst	−97	1,115	185	262
Cambridge Tech.	−90	621	−2	150

Source: MSN MoneyCentral

DiamondCluster International, the firm that lost the least in shareholder value, actually had the second-slowest rate of revenue growth, 99%. The only firm whose revenues shrunk, Cambridge Technology Partners, was also the one whose stock price did the worst. It is clear that size alone does not correlate with stock price performance, since Cambridge Technology is almost twice as large as its next largest competitor, Sapient.

The key factor that has enabled DiamondCluster International and Sapient to lose less value has been their relative lack of dependence on revenue from dot-coms. To gain further insights into the likely survivors of an impending consolidation among Web consultants, it is important to analyze the different strategies of the Web consulting firms. Table 6-3 summarizes the key strategic differences among the seven firms profiled in this chapter. This analysis of Web consultant business strategies suggests that there is some correlation between the Web consulting firm's stock price performance and its strategy. More specifically, the more the Web consultant delivers a closed-loop solution to customers and the less it depends on dot-coms for business, the less the firm's stock price declined in 2000. It appears that the key to a winning stock market performance is the ability of the firm to market its services effectively to CEOs. For example, DiamondCluster International has a marketing organization that does a great job of selling the firm's intellectual capital advantage to CEOs of large companies. Scient is particularly skilled at persuading e-commerce startups' CEOs of Scient's competitive advantages. Firms like Razorfish and Cambridge Technology Partners, which have significant gaps in their delivery capabilities or selling skills, do less well in the stock market.

TABLE 6-3. STRATEGIC DIFFERENCES AMONG SEVEN WEB CONSULTING FIRMS

Company	Competitive Strengths/Weaknesses
DiamondCluster	*Strengths:* Service combines strategy, process, and systems; CEO-focused cross-industry marketing; intellectual capital
	Weaknesses: Potential lack of management bench strength; not using fixed-time/price approach
Proxicom	*Strengths:* Service combines strategy, technology, and creative skills; 800 engagements since 1994; fixed-time/price approach
	Weaknesses: Too dependent on GM and Merrill Lynch as clients; fluctuating financial performance; global organization difficult to manage effectively
Scient	*Strengths:* Service combines strategy and systems; e-commerce client focus; fixed-time/price approach; management depth
	Weaknesses: Dependence on Chase and on e-commerce companies, which could run out of cash; drop in stock price makes retention of staff more difficult; operating losses and high overhead
Sapient	*Strengths:* Service combines strategy, technology, and creative skills; fixed-time/price approach; integration of creative skills

Company	Competitive Strengths/Weaknesses
	Weaknesses: Periodic management changes; difficulty of managing global organization
Viant	*Strengths:* Service combines strategy, technology, and creative skills; fixed-time/price approach; strong client list
	Weaknesses: Too dependent on top clients; choppy financial track record; potential lack of management depth
Razorfish	*Strengths:* Service combines creative and technology; wireless-technology skills; fixed-time/price approach
	Weaknesses: Not as strong on strategy; challenge of integrating acquisitions and managing globally; 1999 loss
MarchFirst	*Strengths:* Strong skills from predecessor, Whittman-Hart; some solid clients
	Weaknesses: Poorly integrated acquisitions; weak financial condition; layoffs hurt morale
Cambridge Technology Partners	*Strengths:* Service combines strategy, technology, and creative skills; fixed-time/price approach
	Weaknesses: Drop in revenues and profits; management turnover; complex organization structure; shareholder lawsuits

Source: Company Reports, Peter S. Cohan & Associates Analysis

The stock market rewards Web consulting firms that are establishing leadership positions in attractive market segments. The market seems to reward firms that are able to conduct effective marketing campaigns targeted at CEOs. Firms such as DiamondCluster International, which have built effective forums for attracting non-dot-com CEOs and helping them to identify the "killer application" for the Web, seem to do quite well among investors. Similarly, firms that have attracted a strong following among leading dot-com CEOs also seem to have earned high returns among investors.

While more quantitative measures of the effectiveness of competitive strategy, such as relative size and rapid revenue growth, are important, the qualitative factors seem to provide greater insight as to which companies investors should avoid. For example, firms like Cambridge Technology Partners, whose revenues shrunk in a market growing at almost 60%, did poorly in the stock market. If an investor sees growth slowing well below the industry average rate, this is a good signal that other investors are likely to sell the stock.

Management Integrity and Adaptability

Management integrity and adaptability are particularly important in Web consulting. Customers and employees are particularly dependent on Web consulting managers' ability to meet their commitments and to anticipate and adapt to changing technology, customer demands, and competitor strategies. If a Web consulting firm loses its ability to meet its commitments to employees or customers, then employees are likely to take a job at a firm that can meet its commitments and customers are likely to take their business elsewhere. Similarly, if a Web consulting firm is unable to adapt effectively to change, then employees will soon find that demand for their services declines and customers will feel that the Web consultant's services are of diminishing value. Since Web consultants depend on talented employees and cutting-edge customers to sustain their market leadership, investors must do their best to scrutinize a firm's management integrity and adaptability.

An analysis of legal proceedings and accounting policies for several Web consultants suggests that not all Web consultants have equal levels of management integrity. As Table 6-4 indicates, two leading Web consulting firms have legal proceedings against them, creating concern about their management integrity. Razorfish is involved in a lawsuit against a former employee alleging fraud. Cambridge Technology Partners is party to 10 class-action lawsuits alleging misrepresentation of the company's financial condition in 1999.

TABLE 6-4. REVENUE ACCOUNTING POLICY AND SHAREHOLDER LAWSUITS FOR SELECTED WEB CONSULTING COMPANIES

Company	Revenue Accounting Policy	Legal Proceedings
DiamondCluster	Recognizes revenue as services are performed in accordance with the terms of the client engagement.	None
Proxicom	When providing services on a time and materials basis, recognizes revenue as it incurs costs. For fixed-price contracts, recognizes revenue using a percentage-of-completion method primarily based on costs incurred.	None

Company Policy	Legal Proceedings	Revenue	Accounting
Scient	Net revenues pursuant to time and materials contracts are generally recognized as services are provided. Net revenues pursuant to fixed-fee type contracts are generally recognized as services are rendered using the percentage-of-completion method of accounting (based on the ratio of costs incurred to total estimated costs).	None	
Sapient	All revenue generated from fixed-price contracts is recognized on the percentage-of-completion method of accounting based on the ratio of costs incurred to total estimated costs. All revenue generated from time and material contracts is recognized as services are provided. Revenues from maintenance agreements are recognized ratably over the terms of the agreements.	None	
Viant	Revenues from fixed-price engagements are recognized using the percentage-of-completion method (based on the ratio of costs incurred to the total estimated project costs).	None	
Razorfish	Recognizes revenues for both time- and materials-based arrangements and fixed-time, fixed-price arrangements on the percentage-of-completion method of accounting based on the ratio of costs incurred to total estimated costs.		In June 1999, a Swedish company, Razor AB, filed a trademark infringement action in Stockholm against Razorfish alleging that the two trade names, Razor and Razorfish, are confusingly similar, and therefore Razorfish was infringing upon Razor's rights under Swedish law.

Company	Revenue Accounting Policy	Legal Proceedings
		Razorfish settled the suit for $86,800. On March 17, 2000, a former consultant to Avalanche Solutions, Inc., a subsidiary of Razorfish, commenced a legal action against Razorfish, alleging fraud and other violations
MarchFirst	Revenues recognized as services are provided.	1998 lawsuit against predecessor, CKS Group, alleging false statement regarding operations.
Cambridge Technology Partners	Revenues from business solutions contracts are recognized primarily on the percentage-of-completion method. The cumulative impact of any revision in estimates of the percent complete is reflected in the period in which the changes become known. Losses on projects in progress are recognized when known. Net revenues exclude reimbursable expenses charged to clients. Revenues from package software evaluation and implementation services are recognized as service is provided, principally on a time and materials basis.	On August 31, 1998, the Company acquired Excell Data Corporation ("Excell"). On November 19, 1998, Excell shareholders filed a lawsuit alleging breach of contract and negligent misrepresentation in connection with the Excell acquisition. On March 2, 2000, the United States District Court for the District of Massachusetts granted Cambridge's motion for summary judgment, dismissing the complaints of the former shareholders of Excell in their entirety. In March and April 1999, certain stockholders of Cambridge filed ten separate class-action lawsuits against Cambridge, alleging misrepresentations and omissions regarding Cambridge's future growth prospects and the progress of Cambridge's reorganization in violation of federal securities laws.

Source: Company Reports

The stock market seems to treat the two Web consulting firms with the legal problems differently from their peers. Razorfish's stock price was a relatively weak performer in 2000, but this weakness cannot be attributed solely to its legal problems. Cambridge Technology Partners' stock price performance was relatively abysmal. This performance can be attributable more directly to the issues raised in the lawsuits alleging misrepresentation and omissions regarding future prospects. With the decline in Cambridge Technology Partners' stock price, employees found that their compensation was much less than they had anticipated. As a consequence, some employees left the firm and started a successful competitor, Breakaway Solutions.

In light of the rapid rate at which the competitive environment changes, the ability of managers to adapt to change effectively is a crucial factor in determining the long-term value of an Internet company. To analyze this factor, we can examine the variance of the percentage change in quarterly earnings. Simply put, if a Web consulting firm is able to increase profits fairly consistently in each quarter, its management is good at adapting to change. Conversely, if a firm's quarterly performance jumps around significantly, its management team is probably less effective at adapting to change. The important point here is that the ability to adapt effectively in a rapidly changing environment can be measured by management's ability to produce fairly consistent financial results.

This measure is particularly useful for analyzing firms that have been publicly traded for a reasonably long period of time. As Figure 6-4 indicates, the firms that have generated the lowest variance in annual percentage change in earnings per share seem to be the ones whose stock prices performed better. Conversely, firms such as Razorfish and Cambridge Technology Partners, which have the highest variance in quarterly percentage change in EPS, had the worst stock market performance.

The ability to manage acquisitions successfully has also been a factor that has differentiated the firms with the steadiest increases in earnings per share from the ones that have had the choppiest performance. In Web consulting, acquisitions are very likely to increase because the low entry barriers make it quite likely that there will be many more firms than the market really needs. In order for these acquisitions to pay off, it will be essential that the acquiring firms pay a low-enough price and do a sufficiently good job of integrating the strong people from the acquired company into the new parent's culture. Sapient has achieved considerable

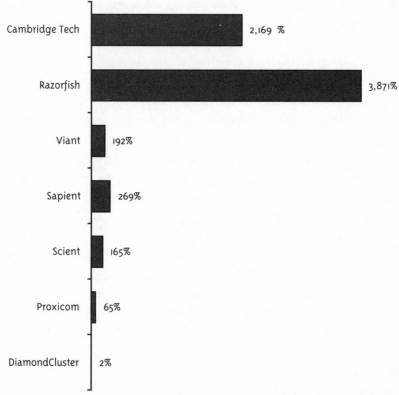

Figure 6-4. Variance in Quarterly Earnings' Percent Change for Selected
Web Consulting Firms, 3/99—3/00
Source: MSN MoneyCentral, Peter S. Cohan & Associates Analysis

success with its acquisitions. It remains to be seen whether avid acquirer MarchFirst will be able to achieve success integrating its many disparate cultures. Its November 2000 decision to lay off 1,000 employees, for example, probably reflects the beginning of more trouble as demand for its services declines.

Management integrity and adaptability are critical factors in differentiating winning Internet companies from their peers. Firms such as DiamondCluster International and Sapient, which have demonstrated consistently high levels of integrity and adaptability, have generated relatively stable earnings growth over time, while firms like Cambridge Technology Partners have suffered from management that lacked the same level of integrity and adaptability—leading to its recent weak stock market performance. Internet investors can use the metrics illustrated here to screen out potentially disappointing management teams and to screen in management teams that are likely to excel.

Brand Family

Since it is often difficult for investors to understand the details of the Web consulting process, investors can analyze the firms' brand families as a gauge of the firms' relative value. The conclusion of the analysis of the different firms' brand families is that there are some important differences among the firms, particularly in terms of their customers and principal investors. These differences foreshadow different shareholder value-generation capabilities and are therefore useful indicators for investors.

An analysis of the firms' customer lists suggests that some of the firms have done a much better job than others at penetrating the most influential reference accounts. Firms that depend heavily on resellers seem to be rewarded less well by investors than firms that depend primarily on a direct sales force. Part of the problem is that resellers demand a piece of the revenue stream of relatively low-margin products, whereas a direct sales force can establish more lasting business relationships with large organizations. These lasting business relationships constitute switching costs that leave more room for higher profit margins.

Firms with the well-known, diversified customer lists are the biggest creators of shareholder value. As Table 6-5 indicates, firms that satisfy these tests include DiamondCluster International, with clients such as non-dot-com leaders AT&T, Fidelity, Microsoft, and Motorola, and Scient, with dot-com clients such as PlanetRx, Sales.com, and Sephora.com. Conversely, Razorfish and Viant, which are more dependent on a smaller number of clients—in a mix of well-known and more obscure firms—tended to perform less well in the stock market. It appears that investors conduct an analysis of the robustness of Web consultants' major clients in order to assess how much revenues and earnings would suffer in the event that a major client switched its business to a competitor or simply cut its budget for Web consulting services. The firms perceived as most likely to survive such a loss seem to have performed better in the stock market. The restructuring initiatives of firms more heavily dependent on dot-coms reflect an effort to shed cash-constrained customers in favor of cash-rich ones.

TABLE 6-5. MAJOR CUSTOMERS OF SELECTED WEB CONSULTING COMPANIES

Company	Major Customers
DiamondCluster	AT&T, Ameritech, Bell Atlantic, Bell South, Carrier Corporation, Fidelity, Lucent, Microsoft, Motorola, and Xerox
Proxicom	Calphalon Corporation, Marriott International, Owens Corning; Harman International, Mercedes-Benz Credit Corp., McKessonHBOC; GE Plastics, Hoffmann-La Roche, and Merrill Lynch
Scient	BenefitPoint, Carstation.com, Chase Manhattan, First Union, Hambrecht & Quist, homestore.com, InnoVentry, Johnson & Johnson, living.com, Miadora, NASDAQ, PlanetRx.com, Sales.com, Sephora.com, Washington Mutual Bank, WineShopper.com
Sapient	United Airlines, BankBoston, AnswerFinancial, Blue Cross/Blue Shield, Goldman Sachs
Viant	American Express, BankBoston, CMGI, Compaq, Deutsche Bank, General Motors, Hewlett-Packard, Kinko's, Lucent Technologies, Polo/Ralph Lauren, LVMH, Radio Shack, Sears, Roebuck, and Sony Pictures Entertainment. In 1998, Viant's five largest clients accounted for approximately 59% of its revenues. During this period, Kinko's, Lucent Technologies, and Compaq each accounted for more than 10% of Viant's revenues. In 1999, Viant's five largest clients accounted for 48% of its revenues. During this period, Compaq and BankBoston each accounted for 10% of Viant's revenues.
Razorfish	Nokia, NatWest, Financial Times, SAP
MarchFirst	Harley-Davidson, Allstate, Williams-Sonoma, Apple Computer
Cambridge Tech.	No data

Source: Company Financial Statements (10Ks), MSN MoneyCentral

While differences among customer bases seem to correlate with differences in stock price performance, a similar correlation does not seem to exist between the quality of a Web consulting firm's investors and its stock price performance. As Table 6-6 indicates, many of the Web consulting firms' shareholders seem to consist of a combination of insiders, large institutional investors, and—in many cases—a small number of venture capital firms. Prestigious venture capital firms (such as Benchmark and Sequoia) own the top creator of shareholder value, Scient. Prestigious venture capital firms (Kleiner Perkins and Mohr, Davidow) also own shares of weaker value creators such as Viant. In short, the quality of the investors in Web consulting firms does not necessarily provide much insight into variations in stock market performance.

TABLE 6-6. MAJOR INVESTORS IN SELECTED WEB CONSULTING COMPANIES

Company	Major Investors
DiamondCluster	Insiders own 15%; three largest institutional owners are Pilgrim Baxter, FMR, and American International Group
Proxicom	General Atlantic Partners, LLC (8.3%), Jeremy Wagner (5.0%), Mellon Financial Corp. (7.2%), FMR Corp. (5.3%)
Scient	Eric Greenberg, chairman (15.7%), Benchmark Capital (14.5%) Sequoia Capital (14.2%), Robert M. Howe, CEO (9.2%)
Sapient	Jerry A. Greenberg (18.0%), J. Stuart Moore (17.7%), Janus Capital Corp. (7.5%), Putnam Investments (7.5%)
Viant	Putnam Investments (26.0%), Mohr, Davidow Ventures (6.9%), Kleiner Perkins Caufield & Byers (5.3%), Robert L. Gett (8.7%)
Razorfish	Spray Ventures AB (31.2%), Omnicom Group (31.4%), Jeffrey A. Dachis (9.2%), Craig M. Kanarick (9.2%)
MarchFirst	Robert Bernard (8.8%)
Cambridge Tech.	Safeguard Scientifics (15.9%), FMR (5.7%), Lord Abbett (5.6%), Massachusetts Financial Services (5.1%)

Source: Company Proxy Statements

Compared to some of the more technically complex Internet business segments, the basic business models of Web consulting firms are more likely to be understood by investors. As a result, the quality of the brand families may not play as important a role for Web consulting as it does for more technically complex Internet business segments such as Internet infrastructure. Investors analyze a Web consulting firm's customers based on a combination of market influence and diversification of revenues. Firms whose revenues are concentrated among a small number of customers, only some of which are market leaders, tend to do less well in the stock market than firms that derive revenues from a larger number of more prestigious customers.

Financial Effectiveness

An important indicator for investors in Web consulting stocks is how well the various firms in the industry are using their financial resources. The industry includes a mix of firms, some growing profitably at a moderate pace, others that are growing revenues very quickly, albeit unprofitably. To compare these firms in terms of financial effectiveness, it makes sense to look at their sales productivity and overhead ratios.

There is a wide variation among the firms in the industry in terms of their sales productivity. As Figure 6-5 suggests, the productivity figures should be analyzed in two groups—those of the startup firms and those of the larger firms. The startup firms actually seem to fall into three sales-productivity tiers. The startup firm with the highest sales productivity is DiamondCluster International, whose sales productivity is almost twice that of its second-tier sales productivity startup peers, Scient and Viant. Investors place a premium on DiamondCluster International' tight financial management, which is evidenced by its much higher sales productivity and the consistent earnings growth we analyzed earlier. A few of the other startup firms, including Proxicom and Razorfish, appear to fall into the bottom tier of sales productivity.

Clearly, the correlation between startup firms' sales productivity and their stock price performance is not tightly correlated. However, as we have noted, DiamondCluster International seems to have pulled away from the pack in terms of its financial management, and investors have rewarded its shares with much higher appreciation than those of its less tightly managed peers.

The absence of a correlation between the SG&A-to-sales ratio and stock price performance is equally pronounced. As Figure 6-6 (page 147) suggests, there is not much correlation between the SG&A-to-sales ratio and relative stock market performance. For example, the worst-

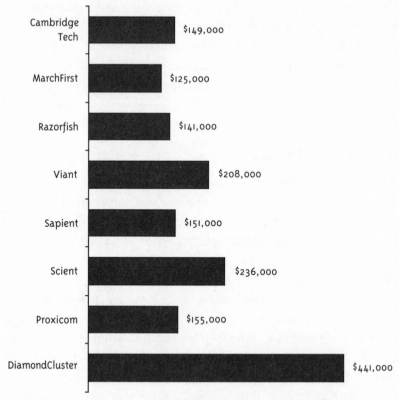

Figure 6-5. Sales per Employee of Selected Web Consulting Firms,
1999
Source: MSN MoneyCentral

performing Web consultant in the stock market, Cambridge Technology Partners, had the lowest ratio of SG&A to sales, while the average performers in the stock market, Scient and Viant, had the highest ratio of SG&A to sales. It is likely that there could be a tighter correlation between the SG&A-to-sales ratio and relative stock market performance in the future if investors' willingness to purchase shares in unprofitable companies continues to diminish. Should investors lose patience with excessive SG&A spending, the less-efficient Web consultants could see weaker stock market performance in the future.

The conclusion from this analysis is that investors have not focused too closely on the financial effectiveness of Web consultants. This general rule seems to have one significant exception: The best stock market per-

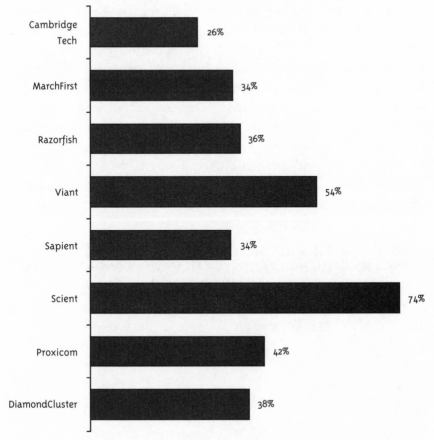

Figure 6-6. Selling, General, and Administrative Expense as a Percent of Sales for Selected Web Consulting Firms, 1999
Source: MSN MoneyCentral

former not only has the tightest control over its quarterly earnings growth but also has the highest sales productivity among its peers by a factor of two. For the remaining competitors, financial effectiveness does not appear to be an important factor in driving variations in the stock market performance of Internet stocks.

Relative Market Valuation

The foregoing analysis of the drivers of Web consultants' stock price performance suggests that the stock market seems to reward recently public

Web consulting firms more highly than older firms making the transition to Web consulting. An analysis of valuation metrics such as price/sales and price/book suggest that the stock market could be mispricing the shares of some of the participants. As Table 6-7 indicates, there are significant differences in the price/sales and price/book ratios of some of the Web consultants whose stock prices have increased at the most rapid rate. DiamondCluster and Sapient—while losing substantial stock market value in 2000—remain more highly valued on a price/sales and price/book basis because they are perceived as likely to survive the dot-com shakeout. Scient and Razorfish, by contrast, appear to be valued as if they could be acquired or closed.

TABLE 6-7. STOCK PRICE AND REVENUE GROWTH PERCENT CHANGE, 2000, AND VALUATION RATIOS FOR EIGHT WEB CONSULTING FIRMS, JANUARY 12, 2001

Company	Stock % Change	Revenue Growth	Price/Sales	(Price/Sales) /Rev. Growth	Price/Book
DiamondCluster Int'l.	−65%	99%	5.60	5.7	3.61
Proxicom	−93	181	1.90	1.0	2.08
Scient	−96	309	0.81	3.8	0.85
Sapient	−83	88	4.91	5.6	3.88
Viant	−92	172	1.52	0.9	0.82
Razorfish	−97	368	0.47	0.1	0.55
MarchFirst	−97	181	0.17	0.1	0.04
Cambridge Tech.	−90	−2	0.63	−31.5	0.24
MEAN	**−89%**	**174%**	**2.00**	**2.5***	**1.51**

Source: MSN MoneyCentral

*Excludes Cambridge Technology Partners

The market valuation metrics also suggest possible mispricing among the publicly traded Web consultants that were formerly client-server consulting firms—Sapient and Cambridge Technology Partners. Sapient may be overvalued, and Cambridge Technology Partners may be undervalued. Sapient's price/sales ratio divided by its revenue growth rate is 5.7. This is by far the largest number among the Web consulting firms analyzed.

While Sapient is the largest of the Web consulting firms analyzed here, its revenue growth rate of 88% falls far below the growth rate of some of its smaller, upstart competitors.

If Cambridge Technology Partners can be turned around, it could be a grossly undervalued stock, with price/sales and price/book ratios well below the valuation benchmarks of other Web consultants. For example, Cambridge Technology Partners' price/sales ratio of 0.6 is more than three times smaller than the average of its peers. The question for investors is whether Cambridge Technology Partners is too sick to be fixed or whether new management could rebuild its franchise. The answer remains to be seen. It is more than likely, however, that the firm's highly paid and less than youthful new management—which hails from railroad company Union Pacific and moribund consulting patriarch Arthur D. Little—will not be able to accelerate the firm's growth. At its low price, a takeover of Cambridge Technology Partners coupled with the injection of new management could add life to this firm's stock.

The relative market valuation analysis raises some interesting investment possibilities. For example, we noted that DiamondCluster International and Cambridge Technology Partners could be undervalued and that Sapient and Scient could be overvalued. In assessing these questions, it is helpful to look back on the relative performance of the firms on the other five indicators we explored in this chapter. For example, do questions about Cambridge Technology Partners' management integrity and adaptability doom it to a continued downward slide in stock price, or could new management be brought in to turn this company around? Has the stock market fully reflected the value of DiamondCluster International's future cash flows into its current price, or is the company significantly undervalued despite its runup? While definitive answers to these questions are beyond the scope of this book, we have described the tools that investors can use to draw their own conclusions.

The Internet Investment Dashboard analysis suggests that Web consulting could be an attractive segment for long-term investors. In Figure 6-7, most of the indicators on the dashboard are dotted. The indicators suggest that investors should approach the Web consulting sector with caution. The industry is large and growing; however, the demise of the dot-coms could slow this growth and put a cap on profitability. On the other hand, many Web consulting firms are offering customers very valuable services. Nevertheless, the industry has low entry barriers, and competi-

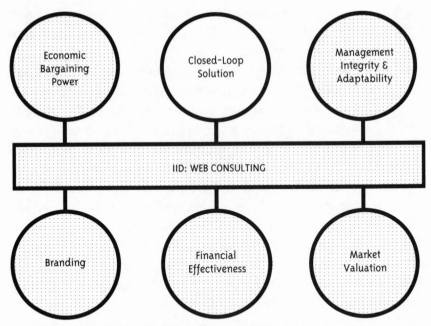

Figure 6-7. Internet Investment Dashboard Assessment of
Web Consulting Segment

Key: Clear = positive performance and prospects; dotted = mixed performance and prospects.

tion is intense. Consolidation is likely as well. As a consequence, investors who can identify the firms likely to be effective consolidators may be able to profit from their investments in Web consulting. There are also significant variations among firms in the industry in their management integrity and adaptability, branding, financial effectiveness, and relative market valuations.

To identify long-term investment opportunities and avoid long-term value destroyers, investors must conduct careful analysis before committing capital to Web consulting. Firms like DiamondCluster International, which excels at marketing to CEOs and has tight financial discipline, may be the best long-term investments—if their valuations do not become too excessive relative to peers. Conversely, firms like Cambridge Technology Partners, which suffers by comparison to its peers in all but the relative market valuation dimension, may be either a good bet to avoid or an undervalued diamond-in-the-rough in need of a turnaround management team. Ultimately, investors will profit by identifying the Web consultants most likely to emerge as leaders following the April 2000 dot-com crash.

PRINCIPLES FOR INVESTING IN WEB CONSULTING STOCKS

Needless to say, the analysis we have conducted here is subject to change. Undoubtedly, some of the strong-performing companies will falter, some of the weak ones could generate positive surprises, and new competitors could emerge to tilt the playing field in a totally different direction. As a result, investors in Web consulting stocks need some principles to guide them through the analytical process as investment conditions evolve.

Web Consulting Investment Principle 1. Monitor developments in new technologies that can put current leaders at a competitive disadvantage. Investors must recognize that the structure of the Web consulting industry is most likely to remain moderately attractive. The most important factor in determining the sector's long-term profitability is the ability of incumbent firms to adapt to important new technologies. For example, if wireless networks become a much more significant platform for U.S. businesses to conduct commerce, then investors will need to assess how quickly incumbent Web consultants will be able to help clients conceive and implement wireless commerce networks. Investors who can anticipate which Web consultants are most likely to profit from new technologies are likely to capture significant shareholder value for their own portfolios.

Web Consulting Investment Principle 2. Analyze the evolution of the upstarts' strategies to assess which will be able to offer closed-loop solutions. Investors should recognize that incumbent firms are more likely to offer customers closed-loop solutions. Investors should assess the quality of these solutions among incumbents and pick the specific firm that offers the most effective closed-loop solution. For example, firms (like Razorfish) that seem to excel in technology and branding and are weaker in strategy may not be able to survive unless they can bolster their strategy consulting skills. Firms that focus on dot-com CEOs, such as Scient, may suffer if funding for dot-com startups withers. This could leave significant opportunity for firms such as DiamondCluster International and Sapient, which have CEO-selling skills and the ability to integrate acquisitions effectively.

Web Consulting Investment Principle 3. Avoid investing in firms where there are questions about management integrity and adaptability. Our analysis indicated questions about management integrity and adaptability in a few firms. This analysis is most useful for investors as a sell signal. If a firm with a track record of avoiding shareholder lawsuits and generating consistent profit growth suddenly deviates from this path, there is a good chance that investors should consider selling the stock. In many cases, such deviations from the path of integrity and adaptability could be a signal that management is no longer up to the job. When these deviations occur, it is likely that management will try to gloss over the problems. Investors may not want to stay along for the ride.

Web Consulting Investment Principle 4. Use compelling brand families as a signal of value in conjunction with other indicators. In the absence of a long financial track record, there are firms, like Juniper Networks, which have such compelling brand families that there is a reasonable chance that they will be able to build a significant business. As new Web consulting firms are taken public, it is likely that there will be some excellent investment opportunities for the investor who can shrewdly assess the quality of these brand families. The best ones may warrant investment, and the worst ones may be worth avoiding.

Web Consulting Investment Principle 5. Employ financial-effectiveness measures if investor patience for high overhead becomes limited. While investors seem to have tolerated the absence of financial discipline among Web consulting firms, this patience could disappear very quickly. One reason for this patience may have been a surfeit of venture capital and the IPO market success of many Web consulting firms. With the postponement of Zefer's IPO in April 2000, however, a new era of investor attitudes was ushered in. It may be that investors will scrutinize more closely the financial effectiveness of Web consulting firms. While this scrutiny could bode well for DiamondCluster International, it could mean trouble for other firms such as Scient, which has a high ratio of SG&A to sales.

Web Consulting Investment Principle 6. Monitor relative valuations most closely within strategic groups to identify potentially under- and overvalued companies. Relative valuations among companies in the same strategic group can help identify specific Web consulting firms whose shares are under- and overvalued. Investors should monitor price/sales, price/sales/sales growth rate, and price-book measures among upstart Web consulting firms and firms that are making the transition to Web consulting from client-server consulting. Investors should analyze differences in valuations among firms in the strategic groups to assess whether the different valuations are consistent with the underlying fundamentals—as illustrated in their relative performance on the other five Internet Investment Indicators.

CONCLUSION

Web consulting could be an attractive area for investors. Because the industry's economic leverage has declined, investors should use the six indicators to pinpoint the firms with good strategies, excellent management, and compelling brand families. Valuation matters for this segment, and depending on the investment time horizon, investors may want to watch closely for opportunities to purchase shares in market-segment leaders during periods of market decline in which excellent companies are punished along with poor ones. Such well-timed investments could yield attractive returns.

Chapter 7

OXYGEN-FREE CAPITALISM: E-COMMERCE STOCK PERFORMANCE DRIVERS

E-COMMERCE HAS STRIPPED the profit out of trade. In that sense, e-commerce is like air without oxygen. Simply put, investors who bet on e-commerce have in most cases found that their investments have dropped in value to the point where it makes sense to sell and take a write-off rather than to rely on the hope of a rebound. Whether the investor is focused on business-to-consumer or business-to-business, the outcome has been pretty much the same. Those who bought at the peak of the frenzy for land-grabbing e-commerce ventures are now sitting on shares whose value has shrunken considerably. The future is uncertain; however, it appears that recovery depends on demonstrable and rapid profit growth. Some firms may achieve this holy grail—most investors should not wait around to find out.

The future for investors in e-commerce stocks is not all doom and gloom. Ariba is an example of where the hope for upside may lie. Ariba's stock rose 400%, from $30 in September 1999 to $150 in September 2000. While at that point the stock was about $30 below its 52-week high, Ariba clearly benefited from its astonishing announcement of second-quarter 2000 results, posting a 578% increase in revenues from the previous year and a net loss that was half the level that analysts had expected. Ariba, whose CEO had been the youngest VP in General Motors' history, was able to do an effective job of posting financial results that swept away the cobwebs of investor worry—replacing them with an irresistible desire to buy Ariba's stock. Despite this good news, investor concern about declining revenue and profit potential caused Ariba stock to plummet to $54 by year end.

By contrast, Amazon.com's stock price performance exemplifies the triumph of cash flow reality over passionate rhetoric from a goofy messenger. Amazon's stock dropped 34%, from $65 in September 1999 to $43 in September 2000. This stock price performance represents something of a disaster and something of a recovery as well. The disaster is that Amazon.com's stock peaked in December 1999 at $113, so its September 2000 $43 price represented a 62% decline. The recovery is that $43 is a 54% leap from Amazon.com's low of $28 during July 2000, after Amazon announced a bigger-than-expected loss and the departure of its chief operating officer. While CEO Jeff Bezos has been the poster child for e-commerce since Amazon.com made it to the cover of the *Wall Street Journal* just prior to its IPO, the fact remains that Amazon.com is in a bunch of low-margin businesses whose margins are driven down as a result of competition from other dot-coms. Amazon's investments in warehouses, customer-service operations, and advertising certainly enable the company to offer customer service and to create a well-known brand. The question for investors is whether Amazon will ever be able to charge enough of a price premium on its products to offset all the investments.

E-commerce means selling products and services over the Web. As we noted in Chapter 2, e-commerce is a relatively small proportion of the entire Internet economy, yet it has received a disproportionate share of attention from the press over the last five years. One of the most likely reasons for the high level of press attention is that e-commerce is relatively easy for most consumers and journalists to understand. Since many consumers and journalists have used e-commerce services, and the basic business models of the industry do not require a deep understanding of technology, journalists have chosen e-commerce as the vehicle for telling the story of the development of Internet business.

This high level of press attention has created a particularly thick layer of verbal fog, which investors must penetrate in order to gain a true picture of the potential risks and rewards of investing in e-commerce stocks. This chapter provides tools that can help investors distinguish between e-commerce sectors that are likely to yield profitable businesses—and stock investments—and those in which managers and investors are more likely to lose money. One of the most important of these tools is an analysis of the industry profit potential of e-commerce vertical market segments.

E-COMMERCE STOCK PERFORMANCE

We begin with an examination of the stock price performance of the e-commerce segment. We noted in Chapter 3 that the stock market performance of a cross section of e-commerce firms was up 201% in 1999. While this average performance compares favorably to NASDAQ's 86% increase in 1999, the 201% increase compares quite unfavorably to the 339% average stock price increase of the companies in our nine-segment Internet stock index in that year.

One important factor behind the lagging performance of e-commerce stocks is the particular timing we have selected for our analysis. More specifically, the e-commerce stocks we analyze in this chapter include a cross section of so-called business-to-consumer (B2C) and business-to-business (B2B) companies. B2C involves selling products and services to consumers. B2B involves selling products and services to businesses. In 1999, the B2C stocks fell out of favor with investors. Prior to 1999, B2C stocks were propelled upward by seemingly limitless enthusiasm for any B2C company, online merchants of beauty products, pet food, etc.

In 1999, investors began to realize that many of these B2C companies were losing huge amounts of money and were unlikely ever to generate positive cash flows. Figure 7-1 summarizes the stock market performance of the eight companies we analyzed. The figure illustrates how the market's disenchantment with B2C led to a significant decline in the value of the stock prices of firms such as eBay, Priceline, and Amazon.com.

The figure also demonstrates that not all B2C companies suffered the same fate in 1999. Specifically, some B2C companies with more robust revenue and cash flows—such as E-Trade and Sportsline—performed rather well in 1999. As we analyze the e-commerce companies, it is useful to bear in mind that investors who follow blanket characterizations such as "all B2C companies lose money and are terrible investments" are likely to miss out on some excellent investment opportunities.

Figure 7-1 also demonstrates the tremendous popularity of firms in the B2B arena; for example, Commerce One and Ariba saw their stock prices soar in 1999. In 2000, however, several of these B2B firms whose stock prices did so well in 1999 saw their stock prices plummet as investors rap-

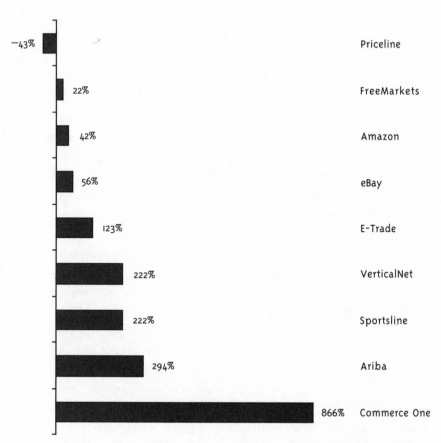

Figure 7-1. Selected E-Commerce Firms' Percent Change in Stock Price,
12/31/98–12/31/99
Source: MSN MoneyCentral

idly abandoned the B2B category. It is worth examining here some of the factors that may have contributed to the stock market popularity of the B2B stocks in 1999 as well as the factors that led to the sector's collapse in 2000.

The primary argument made in favor of B2B during the fall of 1999 was that analysts anticipated that B2B commerce would reach $1.0 trillion in 2003, roughly ten times the anticipated size of B2C. In early 2000, the bloom left the rose when a previous stock market high flyer, FreeMarkets, lost a huge customer, General Motors. Subsequent to dropping FreeMarkets, GM announced that it would partner with Ford and Daimler Chrysler to start their own $250 billion marketplace for purchasing automobile parts in direct competition with the B2B exchanges.

As we will explore later in this chapter, the economic rationale for the explosion in B2B stocks in late 1999 was just as flimsy as the rationale for their collapse in 2000. In order to understand the potential value of B2B, it is important for investors to understand that B2B markets have generally higher switching costs than B2C. Simply put, while consumers may be willing to switch vendors each time they shop for a book or CD, businesses work with other businesses on the basis of many nonprice factors, including timeliness of delivery of the quantity and quality of product ordered, ability to adapt to changing customer needs, financial and management stability, and responsive customer service. These nonprice factors build the costs of switching, which leads to higher margins.

It is also important to understand that the purchasing cooperatives being assembled by GM and its competitors are fraught with significant risks. The U.S. government may stop such purchasing cooperatives from doing business on the grounds that they unfairly exercise monopoly power to pressure suppliers, thereby violating antitrust laws. It is very difficult for fierce competitors in the product markets to cooperate effectively in the market for purchased inputs. There are questions regarding how strong the incentive will be for these competitors to cooperate and overcome operational difficulties in light of the massive decline in the market valuations of B2B firms. Finally, automotive suppliers are likely to resist participating in these purchasing cooperatives owing to fear of seeing their margins cut dramatically and their cost structure exposed.

While B2B stocks performed nicely in 1999, they crashed in 2000 along with many B2C stocks. Figure 7-2 illustrates how the collapse in e-commerce stock prices varied within the sample we examine in this chapter. The most striking aspect of the spring 2000 collapse is that not all e-commerce stocks collapsed equally. eBay, the only profitable company in the group, saw its stock price hold up much better than its peers. In addition, unprofitable firms perceived as category leaders—such as Priceline or Amazon—also held up better, although their declines were more pronounced than eBay's.

The firms that seemed to decline the most were the ones in which investors suddenly lost faith. Firms like Commerce One and Ariba suffered from the announcements we described above from GM and its peers, even though Commerce One and Ariba did not lose revenues directly. Investors' faith was based on the idea that B2B was a larger market and that it had the potential to be profitable. This loss of faith caused investors to dump stocks in the sector with the idea that these firms might become

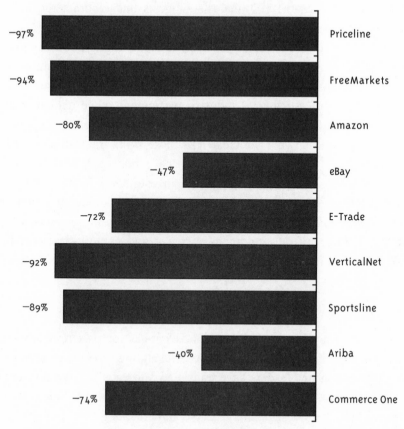

Figure 7-2. Selected E-Commerce Firms' Percent Change in Stock Price,
12/31/99–12/29/00
Source: MSN MoneyCentral

profitable in the future—or show signs of impending profitability—which could make them attractive investments at a lower entry point. Investors also lost faith in firms like Priceline, which subsequently chose to lay off hundreds of employees to conserve limited cash.

INTERNET INVESTMENT DASHBOARD ANALYSIS OF E-COMMERCE

While the foregoing discussion of the relative stock price performance of e-commerce segments is suggestive, it does not offer a fully satisfactory explanation. To obtain such an explanation, we apply the Internet Investment Dashboard (IID) analysis to the e-commerce segment.

As we will see, the IID analysis suggests that e-commerce is not likely to generate attractive investment returns in the future. The e-commerce industry has virtually no economic bargaining power, as illustrated by its negative margins. While many participants offer customers a closed-loop solution, these providers are only able to gain a share of an unprofitable business. Several industry participants have demonstrated their ability to adapt effectively to change and have created a compelling brand family, but it is not yet clear how these skills translate into profitability. Furthermore, most participants have managed their financial resources based on the assumption that investors would be willing to finance their operating losses. As investor sentiment suddenly shifted in April 2000, bankruptcies and shotgun mergers left only a few leaders standing—albeit still struggling to conceive and exploit strategies for profitable operation.

Economic Bargaining Power

E-commerce is a potentially huge business that has yet to exercise any significant economic bargaining power. As a result, the growing revenues of e-commerce industry leaders have translated into equally large operating losses. The e-commerce market actually consists of at least three distinct segments. The largest segment, B2B e-commerce, is growing at an 85% compound annual rate and is anticipated to reach $1.3 trillion by 2003, as illustrated in Figure 7-3. Firms such as Commerce One, Ariba, and VerticalNet are anticipated to derive significant shares of their revenues from this market.

Another very large segment of e-commerce is online auctions. As illustrated in Figure 7-4 (page 162), the value of goods sold in online auctions is anticipated to reach $52.6 billion by 2002, growing at a compound annual rate of 56%. eBay has clearly established itself as the leader in this segment. While eBay faces competition from Yahoo and Amazon, online auctions constitute a business in which first-mover advantages are significant, particularly if they can be sustained by improving security and reliability.

A third segment, which has gained by far the lion's share of press attention over the last several years, is B2C. As Figure 7-5 (page 162) illustrates, B2C is expected to generate revenues of $145 billion in 2003, growing at a 42% annual rate and representing about 10% of the overall e-commerce market in 2003. In short, B2C is likely to be a much smaller

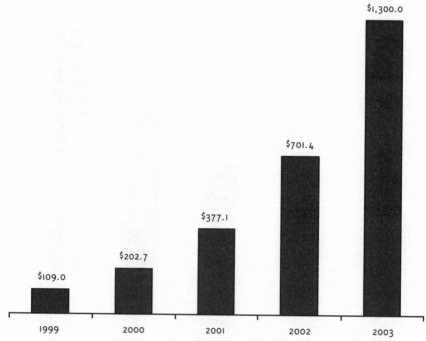

Figure 7-3. Business-to-Business Electronic Commerce, 1999–2003, in Billions of Dollars

Source: International Data Corp.

market than B2B and to grow a bit more slowly. B2C market leaders include the relatively large (and unprofitable) Amazon and Priceline, to name two from our sample of companies.

It is worth noting that B2C and B2B are aggregate categories that mask significant differences in size among specific vertical market segments. As Figure 7-6 (page 163) illustrates, by far the largest B2C vertical market segment is leisure travel, at an estimated $14 billion in 2000 revenues, while the smallest category is automobiles, with an estimated $400 million in 2000 revenues. In general, consumers appear most willing to purchase items online where they feel a limited need to touch and feel the item being purchased before making the buying decision and where the dollar amount of the purchase is relatively low. Conversely, when buyers need physical contact with the item being purchased and the dollar amount of the purchase is significant, as is the case with automobiles, the magnitude of the vertical market tends to be smaller. Similar differences exist among different B2B vertical market segments.

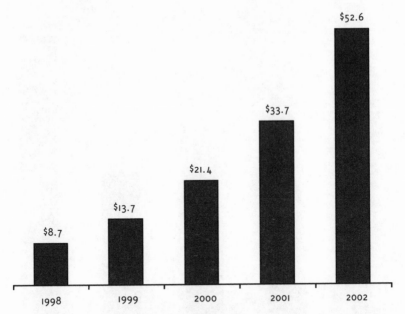

Figure 7-4. Value of Goods Sold in Online Auctions, 1998–2002,
in Billions of Dollars

Source: Forrester Research

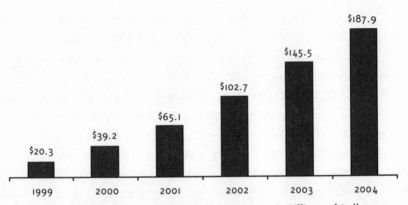

Figure 7-5. Total U.S. B2C Revenues, 1999–2004, in Billions of Dollars

Source: Forrester Research

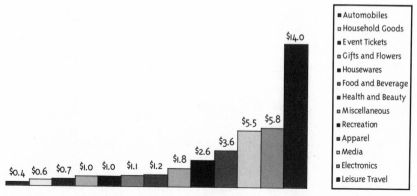

Figure 7-6. B2C Revenues by Product Category, 2000, in Billions of Dollars
Source: Forrester Research

While the magnitude of e-commerce appears impressive, it is important to note, as we did in Chapter 2, that the $202 billion B2B e-commerce market estimate represents about 2% of U.S. Gross Domestic Product. While this percentage is forecast to grow dramatically in the next several years, it remains a fairly small proportion of all commerce—a proportion that is much smaller than the amount of press attention it has been receiving over the last several years.

One aspect of e-commerce that has received significant attention has been its almost complete absence of profitability. As Table 7-1 illustrates, the average net margin for the nine e-commerce companies sampled in this chapter is −145%. This means that for every dollar that these companies generated in revenue, they lost $1.45. This statistic speaks eloquently to the lack of economic bargaining power of the average e-commerce company.

TABLE 7-1. MARGINS FOR NINE E-COMMERCE FIRMS, DECEMBER 2000

Company	Stock % Change	Gross Margin	Operating Margin	Net Margin	2-Year Trend
Commerce One	−74%	83%	−76%	−78%	+
Ariba	−40	86	−391	−391	−
Sportsline	−89	54	−502	−502	−
VerticalNet	−92	86	−101	−101	−−
E-Trade	−72	−45	−18	−6	++

Company	Stock % Change	Gross Margin	Operating Margin	Net Margin	2-Year Trend
eBay	−47	92	22	9	+
Amazon.com	−80	26	−37	−37	+
FreeMarkets	−94	47	−105	−105	−−
Priceline	−29	24	−96	−96	−−
MEAN	**−76%**	**50%**	**−145%**	**−145%**	−

Source: MSN MoneyCentral

Note: Two-year trends represent the extent to which net margins have changed in the most recent two years. ++ signifies a large improvement, + indicates a small improvement, − indicates small deterioration, −− indicates large deterioration.

The statistic also masks wide variations in the profit margins of the specific firms sampled. For example, eBay, with its 9% net margin, is the only firm that has actually generated net profits fairly consistently since it went public. eBay has been profitable because it has created a unique way of transacting business that involves taking fees for listing items for sale and earning a commission on closed transactions. Because eBay is the largest online auction site, much of its growth occurs because its market leadership attracts new buyers and new sellers at the expense of its competitors' growth.

By contrast, VerticalNet loses $1.01 for every dollar of revenue it generates. VerticalNet's historical business model has involved attracting advertisers to its collection of fairly disparate vertical market segments' trade magazines. In fact, VerticalNet's advertising-based revenue model takes a weak business and makes it less profitable because VerticalNet must spend so much money to encourage advertisers to spend. While e-commerce is considered the core of its business, VerticalNet has yet to generate significant e-commerce revenue.

While the details of each firm's business model are not crucial to understand here, it is important for investors to recognize what factors are driving the lack of economic bargaining power of the average e-commerce company. Economic bargaining power is important because it enables a firm to charge a price high enough to cover its costs and to earn profits so that it can reinvest to grow its business. Most e-commerce companies have yet to attain this power. In B2C, as we noted earlier, e-commerce firms have been selling products with inherently low profit margins at a discount with the intent of "making it up on volume." The thin gross margins from selling items like books and CDs (for example, Amazon.com's gross margin is 26%) are not sufficient to finance significant spending on sales, marketing, and advertising and on building warehouses to store the products and customer-service operations to build long-term relationships with customers. Up until the fall of 1999, B2C firms were successful at persuading investors to trade their cash for equity in this business model, but they could not convince customers to pay for all these investments.

The e-commerce segment thus has low economic bargaining power. While the segment is anticipated to become large over the next three years, the profit margins remain quite negative. Firms are competing intensely for category dominance. This competition entails significant investments, including low product prices, aggressive sales, marketing, and advertising, millions of square feet of warehouses filled with inventory, and calling centers filled with customer-service staff. For much of 1999, e-commerce firms were successful at persuading investors to part with their cash in exchange for equity in the future cash flows expected to flow from e-commerce. Currently it appears that investors have lost much of their confidence in this positive-cash-flow future, but customers continue to expect e-commerce to deliver extra convenience at a lower price. This is a prescription for bankruptcy.

Closed-Loop Solution

One thing that many e-commerce firms have done successfully is to offer customers a closed-loop solution. For example, a visitor to Amazon.com can identify a category of books, peruse comments of other readers and reviewers, place an order, and expect delivery often within two or three days—generally at a price that is very competitive. In short, many e-commerce firms have figured out how to offer customers added convenience and better service than land-based retailers, all at a lower price. This enhanced value proposition is a prescription for rapid revenue growth.

One way to assess the relative effectiveness of the strategies of firms in the e-commerce market is to analyze the firms' relative size, revenue growth, and market capitalization. This analysis can help us gain insights into which firms are generating the greatest product market momentum and the extent to which that momentum has translated into greater stock price appreciation.

The correlation between momentum in the product markets and the capital markets is partially borne out by the evidence. As Table 7-2 indicates, it appears that a firm's rate of shareholder value creation is somewhat correlated with a combination of its e-commerce category (e.g., B2B, B2C, or online auctions), its rate of revenue growth, and its relative size.

TABLE 7-2. REVENUE MOMENTUM OF NINE E-COMMERCE FIRMS, JANUARY 12, 2001, IN MILLIONS OF DOLLARS

Company	Stock % Change	12-Months Revenue	Revenue Growth	Market Capitalization
Commerce One	−74%	$228	1,157%	$4,770
Ariba	−40	279	515	10,800
Sportsline	−89	95	88	145
VerticalNet	−92	165	1346	517
E-Trade	−72	2,202	232	3,700
eBay	−47	371	97	11,000
Amazon.com	−80	2,466	86	6,050
FreeMarkets	−94	65	338	812
Priceline	−97	1,176	222	437

Source: MSN MoneyCentral

Comparing firms within B2B, the fastest-growing firm—Commerce One—saw its revenues appreciate 1,157% and its stock price drop 74% in 2000. Ariba, a larger firm in competition with Commerce One, saw its revenues grow a relatively paltry 515% while its stock price fell a relatively low 40%. VerticalNet, a much smaller company, experienced a much faster 1,346% revenue growth, but its stock price plunged 92%.

While this logic seemed to hold up for firms in the B2B sector, it fell apart in the B2C arena, where a firm's relative size and profitability seemed to play a more important role in driving relative stock price performance. For example, Priceline's revenues increased over 500%, but the company lost so much money ($2.19 for every dollar of revenue) that investors cut 97% from Priceline stock in 2000. Amazon, a much larger firm in terms of revenues, experienced a relatively slow 96% growth in revenues and lost 80% of its value in 2000.

In order to gain meaningful insights that can explain how the differences in strategies of the firms translate into different levels of return in the stock market, it is important to analyze the different strategies of the e-commerce firms. Table 7-3 summarizes the key strategic differences among the nine firms profiled in this chapter.

TABLE 7-3. STRATEGIC DIFFERENCES AMONG NINE E-COMMERCE FIRMS

Company	Competitive Strengths/Weaknesses
Commerce One	*Strengths:* Closed-loop solution includes procurement software, purchasing Web site, content management, order availability information, status tracking, and transaction support; strong partnerships and customers *Weaknesses:* Substantial operating losses
Ariba	*Strengths:* Strong customer list includes DuPont, Federal Express, Chevron, and Hewlett-Packard; excellent management *Weaknesses:* Challenge to integrate acquisitions; substantial operating losses; need better supply chain capabilities
Sportsline	*Strengths:* Excellent sports content; access to CBS-TV advertising in exchange for CBS ownership stake *Weaknesses:* Excessive dependence on advertising revenue; weak financial position
VerticalNet	*Strengths:* 50 Web sites with recently added online auction capability; backing from Internet Capital and Microsoft *Weaknesses:* Lack of industry focus makes it difficult to compete with single-industry competitors; weak financial performance; management turnover
E-Trade	*Strengths:* 2 million accounts; operates globally; strong financial backing from Softbank *Weaknesses:* Overtaken by Schwab (now occupies number three position); management style is too internally focused; losses
eBay	*Strengths:* 4,000 merchandise categories and 10 million registered users; leading market share creates virtuous cycle *Weaknesses:* Track record of site crashes; challenges in dealing with fraudulent auctions

Company	Competitive Strengths/Weaknesses
Amazon.com	*Strengths:* Sells millions of books, CDs, DVDs, videos, toys, tools, electronics, and conducts auctions; leading market share
	Weaknesses: Thin margins; huge capital requirements; unclear path to profitability
FreeMarkets	*Strengths:* Strong customer list; expertise in chemical and coal markets
	Weaknesses: Challenge of collecting transaction fees; weak financial performance; insufficient vertical market focus
Priceline	*Strengths:* Innovator of reverse auctions for airline tickets, hotel rooms, cars, home mortgages, refinancing, equity loans, grocery products, and gasoline
	Weaknesses: Potentially shaky accounting; management turnover; unclear patent protection; huge operating losses

Source: Company Reports, Peter S. Cohan & Associates Analysis

The analysis of e-commerce business strategies in Table 7-3 suggests that there is some correlation between the e-commerce firm's stock price performance and the robustness of its competitive strategy. This correlation appears particularly strong in the B2B sector. For example, Commerce One appears to have been the most successful at putting together for its customers a closed-loop solution that includes procurement software, Web site, content management, order availability information, status tracking, and transaction support. Ariba, with an excellent customer list for its software, has been acquiring its way into the full suite of capabilities that Commerce One appears to offer to its customers. VerticalNet is the furthest behind, having only recently added an online auction capability.

The relative state of strategic development seems to correlate well with relative stock price performance. More specifically, Commerce One's stock price went up the most in 1999, followed by Ariba's, then, much farther behind, by VerticalNet's price. The relative stock price performance for these firms seems to follow from their competitive positions.

B2C investors also seem to relate stock price performance to a firm's competitive strengths and weaknesses. An interesting example of this is eBay, whose financial performance has been relatively consistent and profitable. Despite its market leadership in the online auction business, in 1999 investors seemed to be punishing eBay for its less-than-perfect operations. For example, eBay's site crashed on a number of occasions during 1999; eBay also contended with several fraudulent auctions. While some may have argued in the company's defense that it was simply experiencing some of the challenges of rapid growth, it may be that investors were less forgiving of eBay's problems in handing the company a 47% decline in its shares.

As we have stated, the stock market rewards e-commerce firms that have the most robust competitive strategies. The market seems to reward firms that have combined all the capabilities that customers need to experience the full value-creating potential of e-commerce. Firms whose capabilities are still evolving to meet customer needs experience less stock price appreciation than those firms with solidly established capabilities. However, investors even punish firms, such as eBay, that have achieved market leadership but that are unable to sustain consistently high levels of customer service.

While more quantitative measures of the effectiveness of such competitive strategies as relative size and rapid revenue growth are important, the quantitative factors seem to provide greater insight as to which companies investors should avoid. For example, firms like Priceline, whose capital position and cash-burn rate raise questions about its ability to survive, did particularly poorly in the stock market. If an investor sees an e-commerce firm's cash-burn rate rapidly approaching (and soon exceeding) its available cash, this is a good signal that other investors are likely to sell the stock (and the company is soon likely to perish).

Management Integrity and Adaptability

Management integrity and adaptability are particularly important in e-commerce. Customers and employees are particularly dependent on e-commerce managers' ability to meet their commitments and to anticipate and adapt to changing technology, customer demands, and competitor strategies. If an e-commerce firm loses its ability to meet its commitment to employees or customers, then employees are likely to take a job at a firm that can meet its commitments and customers are likely to take their business elsewhere.

An analysis of legal proceedings and accounting policies for several e-commerce firms suggests that not all firms have equal levels of management integrity. As Table 7-4 indicates, several e-commerce firms, such as

Amazon and Priceline, seem to employ accounting techniques that, while apparently consistent with generally accepted accounting principles, appear to push the envelope with the intent of increasing reported revenues and—since stock market valuations are keyed off of revenues—to inflate market valuations as well. Furthermore, several e-commerce firms in our sample of companies are involved in legal matters which, while currently unresolved, raise questions regarding management's business practices.

TABLE 7-4. REVENUE ACCOUNTING POLICY AND SHAREHOLDER LAWSUITS FOR SELECTED E-COMMERCE COMPANIES

Company	Revenue Accounting Policy	Legal Proceedings
Commerce One	Commerce One recognizes revenues from license agreements on delivery and acceptance of the software if there is persuasive evidence of an arrangement, collection is probable, the fee is fixed.	None
Ariba	Revenue allocated to software licenses is generally recognized upon delivery of the products.	None
Sportsline	Barter transactions accounted for 19% and 17% of total revenue for the nine months ended September 30, 1999 and 1998, respectively.	None
VerticalNet	All advertising revenues are recognized ratably over the period in which the advertisement is displayed, provided that collection is reasonably assured. VerticalNet reflects gross revenues and related product costs of exchange transactions in its consolidated financial statements since NECX takes title to the products exchanged in such transactions and is exposed to both inventory and credit risk related to the execution of the transactions.	None

Company	Revenue Accounting Policy	Legal Proceedings
E-Trade	E-Trade derives transaction revenues from commissions related to retail customer broker-dealer transactions in equity and debt securities, options and, to a lesser extent, payments from other broker-dealers for order flow.	Numerous lawsuits alleging false and deceptive advertising and other communications regarding E-Trade's commission rates and ability to timely execute and confirm online transactions; damages arising from alleged problems in accessing brokerage accounts and placing orders; damages arising from system interruptions, including those occurring in February 1999; and unfair business practices regarding the extent to which initial public offering shares are made available to E-Trade's customers.
eBay	Online transaction revenues are derived primarily from placement fees charged for the listing of items on the eBay service and success fees calculated as a percentage of the final sales transaction value. Revenues related to placement fees are recognized at the time the item is listed, while those related to success fees are recognized at the time that the transaction is successfully concluded.	On March 23, 1999, eBay was sued by Network Engineering Software, Inc. ("NES") in the U.S. District Court for the Northern District of California for its alleged willful and deliberate violation of a patent. On September 1, 1999, eBay was served with a lawsuit filed by Randall Stoner alleging that the eBay listed "bootleg" or "pirate" recordings.
Amazon.com	Amazon.com recognizes revenue from product sales, net of any promotional gift certificates, when the products are shipped to customers, which is also when title passes to customers.	In January and February 2000, three federal class action lawsuits were filed against Amazon and its subsidiary, Alexa Internet. The lawsuits allege that Alexa Internet's tracking and storage of Internet Web usage paths violate federal and state statues prohibiting computer

Company	Revenue Accounting Policy	Legal Proceedings
	Outbound shipping charges are included in net sales and amounted to $239.0 million, $94.1 million, and $24.8 million in 1999, 1998, and 1997, respectively.	fraud, unfair competition, and unauthorized interception of private electronic communications, as well as common law proscriptions against trespass and invasion of privacy.
FreeMarkets	Recognizes revenues from fixed fees as FreeMarkets provides service.	Ten securities fraud class-action suits.
Priceline	The manner in which and time at which revenues are recognized differs depending on the product or service sold through the Priceline.com service. With respect to airline ticket, hotel room, and rental car services, revenues are generated by transactions with customers who make offers to purchase airline tickets and reserve hotel rooms and rental cars supplied to priceline.com by participating sellers. Revenues and related costs are recognized if, and when, Priceline.com accepts and fulfills the customer's offer. Because Priceline.com is the merchant of record in these transactions, revenue for these services includes the offer price paid by the customer, net of certain taxes and fees.	Marketel's suit alleges patent infringement of business model, which allegedly was provided in confidence approximately ten years ago.

Source: Company 10Ks

The accounting practices of Amazon and Priceline suggest that certain amounts are counted as revenues that, while they may be in accordance with generally accepted accounting principles, do not seem consistent with common sense. For example, Amazon.com counts outbound shipping charges as part of revenue. In 1999, Amazon.com generated outbound shipping charges of $239 million. While this amount represented a little over 10% of revenues, it simply does not seem consistent with common sense to include shipping costs as revenues. Simply put, it would make more sense to include the shipping charges as an expense that is reimbursed by customers rather than as part of Amazon's sales. By including this amount in revenue, Amazon grosses up its revenue, which may have contributed to a higher valuation.

Priceline includes as revenues the total price of the tickets that customers purchase through its system. While Priceline claims that it is the merchant of record for these transactions, the fact remains that as an economic matter Priceline is merely a broker, and the actual airline services are delivered by the seller of the ticket—the airline. Priceline's slim 24% gross margins suggest that including the price of the ticket as revenue, while it may be consistent with generally accepted accounting principles, actually inflates Priceline's revenues. It may make a bit more sense to use Priceline's much lower gross profit as its revenue figure. Such an adjusted revenue amount would probably give Priceline a significant valuation haircut.

As noted earlier, many of the e-commerce companies covered in this chapter have pending lawsuits. The lawsuits are a rogues' gallery of the ethical and regulatory problems facing Internet companies. For example, Priceline and eBay were sued for patent infringement. E-Trade was sued for a variety of alleged wrongs, including misleading advertising, poor trade execution, and unfair terms of distributing IPO shares. Amazon, through its Alexa Internet subsidiary, was sued for violation of privacy.

While these lawsuits and accounting practices do not necessarily mean that the e-commerce firms violated any laws, they do suggest that there was some bending of limits. This bending of limits could constitute a violation of investor and customer trust in some cases. The impact of this environment of diminished trust could contribute to an unwillingness on the part of investors to make a bet on an uncertain commercial future.

In light of the rapid rate at which the competitive environment changes, the ability of managers to adapt to this change effectively is a crucial factor in determining the long-term value of an Internet company. To analyze this factor, we can examine the variance of the percentage change in quarterly earnings. Simply put, if an e-commerce firm is able to increase profits fairly consistently in each quarter, its management is good at adapting to change. Conversely, if a firm's quarterly performance jumps around significantly, its management team is probably less effective at adapting to change. The important point here is that the ability to adapt effectively to change in a rapidly changing environment can be measured by management's ability to produce fairly consistent financial results.

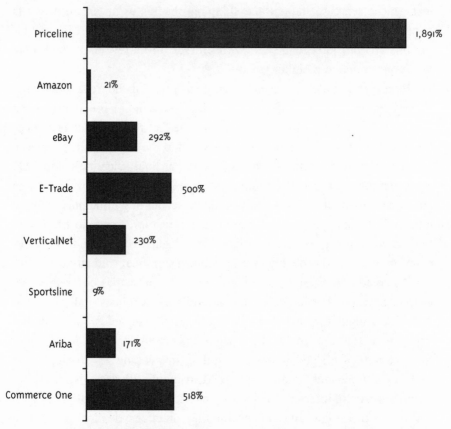

Figure 7-7. Variance in Quarterly Earnings' Percent Change for
Selected E-Commerce Firms, 3/99–3/00
Source: MSN MoneyCentral, Peter S. Cohan & Associates Analysis

In the case of e-commerce firms, it is clear that this measure—a useful one for many other Internet business segments—is not particularly useful. As Figure 7-7 indicates, consistent earnings growth is not a relevant measure for e-commerce firms in predicting stock price appreciation. The fundamental reason for this is that earnings have not been perceived as an important measure for e-commerce firms. For much of 1999, e-commerce companies generated huge losses, yet investors wanted to own them and save them significant market capitalizations. Ironically, the firm with the most consistent performance (consistent levels of net losses) actually had the worst stock market performance.

Management integrity and adaptability are critical factors in differentiating winning Internet companies from their peers. The average e-commerce company analyzed here seems to have achieved a less-than-outstanding performance on both of these qualitative factors. Perhaps Commerce One and Ariba have demonstrated an ability to adapt effectively in terms of building or acquiring the skills that customers require. It remains to be seen, however, whether these apparent adaptations will result in sustainable profitability.

Brand Family

Since it is often difficult for investors to understand the details of e-commerce process, investors may find it useful to analyze the firms' brand families as a gauge of the firms' relative value (see Table 7-5). The conclusion of the analysis of the different firms' brand families is that there are some important differences among the firms, particularly in terms of their customers and principal investors. These differences foreshadow different capabilities of shareholder-value generation and are therefore useful indicators for investors.

The brand families of e-commerce companies vary significantly, depending on whether the firms are B2B or B2C. B2B firms can be evaluated on the basis of the quality of their partners and their corporate customers. Based on these criteria, both Commerce One and Ariba have assembled excellent reference customers and well-regarded partners. These customers and partners have significant strategic value to these firms because these names make it easier to make new sales to other customers. Firms that put themselves in good company are likely to attract more good company. The effect of this virtuous cycle for investors is to create a reasonable expectation of future growth.

For B2C firms, the brand cannot be established on the basis of the use of reference customers, since individuals are not likely to be influenced by other relatively anonymous individuals in a significant way. If a celebrity is associated with a particular Web site, however, that celebrity can contribute to the growth of the site. For example, Sportsline undoubtedly benefits from its association with Tiger Woods, and it is likely that Priceline may have benefited from its use of *Star Trek*'s William Shatner as a spokesman. It appears that the firms that performed best in 1999's markets did benefit from having well-known customers and partners. As Priceline's 2000 stock market performance suggests, cash-flow problems outweigh the marketing benefits of being associated with these personalities.

TABLE 7-5. MAJOR CUSTOMERS AND PARTNERS OF SELECTED E-COMMERCE COMPANIES

Company	Major Customers/Partners
Commerce One	Partners include Appnet, Wells Fargo, PeopleSoft, Compaq, Bellsouth, and the Sabre Group. Customers include Bass, Pacific Gas & Electric, Singapore Telecom, and British Telecom.
Ariba	Partners include Harbinger, Impresse, Andersen Consulting, Oracle, PeopleSoft, and SAP. Customers include Cadence, Federal Express, Sonoco, MCI WorldCom, Nestlé, and General Motors.
Sportsline	Web sites (sf49ers.com, worldseries.com), sports personalities (tigerwoods.com, shaq.com), and sports odds (vegasinsider.com). Viacom owns 18% of the company since its acquisition of CBS.
VerticalNet	Partners include AltaVista, Excite, Yahoo, Tradex Technologies, Deja.com, Visio, and Powerize.
E-Trade	Customers are individuals.
eBay	In February announced an agreement with Disney to develop cobranded consumer-to-consumer and merchant-to-consumer Web sites, including merchant-to-consumers sites for Disney.com, ESPN.com, and ABC.com; also announced eBay Japan, a joint venture with NEC that will offer an auction Web site in Japanese. In March announced Wells Fargo will take an equity stake in eBay's subsidiary, Billpoint, which provides secure credit card payment services over the Internet. Also in March, launched Ebay Business Exchange Service, which allows small businesses to buy and sell items including computer hardware, software, electronics, and office equipment. Other partners include E-Stamp, iShip, Kinko's, and Skytel Communications.
Amazon.com	Sothebys.Amazon.com, which offers items in categories including jewelry, watches, silver, furniture, entertainment, sports memorabilia, fashion, coins, paintings, and photographs. In February, launched Toolcrib.Amazon.com, an online retailer of tools and equipment for professionals. Partners include America Online, Yahoo, and Red Brick.
FreeMarkets	United Technologies, Quaker Oats, Owens Corning, Eaton, and Emerson Electric.
Priceline	William Shatner (terminated in 2001).

Source: *Red Herring*, Microsoft MoneyCentral, Company Financial Statements (10Ks)

Differences among customer bases and partners seem to correlate with differences in stock price performance. A similar correlation seems to exist between the quality of an e-commerce firm's investors and its stock price performance. As Table 7-6 indicates, many of the top stock-market-performing e-commerce firms' shareholders seem to consist of a combination of insiders, large institutional investors, and—in many cases—a small number of venture capital firms. The best performers, such as Ariba, Sportsline, and E-Trade, seem to benefit from the investment support of such prestigious investors as Crosspoint Ventures, CBS, and Softbank. Interestingly, some of the weaker-performing e-commerce firms seem to be owned by insiders coupled with somewhat less prestigious venture capital firms and institutional investors.

TABLE 7-6. MAJOR INVESTORS IN SELECTED E-COMMERCE COMPANIES

Company	Major Investors
Commerce One	Insiders own 31%, institutions own 22%
Ariba	Crosspoint Venture Partners (5.4%), Amerindo Investment Advisors (5.7%), Keith J. Krach (11.1%)
Sportsline	CBS Corporation (19.8%), Massachusetts Financial Services (10.2%), Putnam Investments (7.1%), MediaOne Interactive Services (6.1%), Michael Levy (4.8%)
VerticalNet	Internet Capital Group (33.4%), Michael J. Hagan (3.1%), Mark L. Walsh (2.0%)
E-Trade	Christos M. Cotsakos (3.0%), William A. Porter (3.3%) Softbank Holdings (26.1%), General Atlantic Partners (3.2%)
eBay	Pierre M. Omidyar (27.0%), Jeffrey S. Skoll (16.0%), Margaret C. Whitman (4.9%)
Amazon.com	Jeffrey P. Bezos (33.62%), Janus Capital Corporation (10.48%), Thomas H. Bailey (10.48%)
FreeMarkets	Glenn Meakem (10%), CSM Partners (6.5%), Goldman Sachs (6.2%), Sam Kinney (6.0%)
Priceline	Jay S. Walker (37.35%), Richard S. Braddock (9.97%), William E. Ford (12.06%), General Atlantic Partners (12.06%), Paul G. Allen (5.34%)

Source: Company Proxy Statements

It appears that investors have paid a significant amount of attention to brand family in picking e-commerce firms in which to invest. The firms with the best customers, partners, and investors performed better in the stock market in 1999 than firms that did not do as well in accumulating a compelling brand family. It is possible that this difference can be accounted for by the faith that investors place in well-established brand names providing an element of stability in a rapidly changing world in which commercial outcomes are so uncertain.

Financial Effectiveness

Prior to the second quarter of 2000, investors did not use financial effectiveness as a means of discriminating among e-commerce firms. While there is no correlation between e-commerce firms' financial effectiveness and their stock market performance, the analyses of sales productivity and operational efficiency highlight some interesting differences among different e-commerce business models.

There is a wide variation among three classes of e-commerce firms in terms of their sales productivity. As Figure 7-8 suggests, the productivity figures should be analyzed in three groups—truly innovative e-commerce business models, financial services e-commerce businesses, and all the rest. The truly innovative businesses—eBay and Priceline—are considered innovative because it is difficult to imagine how these businesses could operate without the Internet. For eBay and Priceline, it is interesting to point out that sales per employee are much higher than for other e-commerce businesses. The ability to achieve much higher sales per employee actually suggests some extraordinary profit potential in the future, since it appears that sales could become quite enormous without adding a proportional number of additional people. Simply put, innovative e-commerce business models can scale very efficiently. As we noted earlier, however, such rapid scaling did not translate into rapid stock price appreciation in 1999.

The one financial-services business in our sample, E-Trade, had a significantly higher-than-average level of sales per employee, although it was not close to the levels achieved by eBay and Priceline. The other e-commerce companies in the sample had relatively low sales productivity. This low sales productivity is particularly notable for the e-commerce firms that are dependent on a sales force to close corporate sales. It is worth pointing out that sales forces for these firms must persuade potential customers to change their

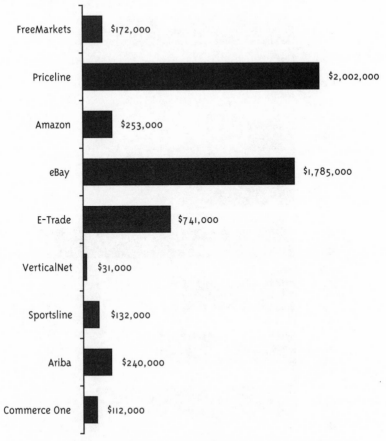

Figure 7-8. Sales per Employee of Selected E-Commerce Firms, 1999
Source: MSN MoneyCentral

business practices before they can convince these customers to use their particular service to enable those changes. The consequence of this challenge from a sales-productivity perspective is that B2B e-commerce firms run a real danger of having very low sales productivity as long as they depend on long sales cycles to close deals.

The absence of a correlation between the SG&A-to-sales ratio and stock price performance is equally pronounced. As Figure 7-9 suggests, there is not much of a correlation between SG&A to sales and relative stock market performance. For example, the worst-performing e-commerce firm in our stock market sample, eBay, had a ratio of SG&A to sales that was less than half the level of the best-performing firm in the market, Commerce One. Should investors lose patience with excessive SG&A

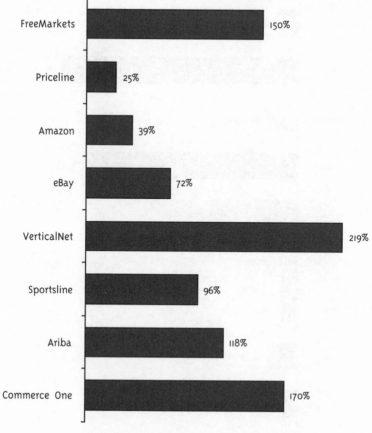

Figure 7-9. Selling, General, and Administrative Expense as a Percent of
Sales for Selected E-Commerce Firms, 1999
Source: MSN MoneyCentral

spending, the less efficient e-commerce firms could see weaker stock market performance in the future.

Although it is apparent that investors did not focus too closely on financial effectiveness among e-commerce firms in 1999, this general rule seems to be changing. As investors abandon e-commerce stocks, it is likely that only the firms that can generate funds from internal operations will be able to stay in business. As a result, e-commerce firms that focus intensely on financial effectiveness are more likely to survive in the future. Investors can then use financial effectiveness as a useful means of discriminating between the quick and the dead. The sooner they can identify e-commerce firms that are likely to run out of cash, the sooner they can dump their shares in these companies and shift the funds thus freed to more promising investments.

Relative Market Valuation

The stock market's approach to valuing e-commerce companies suggests that different sectors are more out of whack than others. As Table 7-7 indicates, on the basis of price/sales ratios, for example, the firms in our sample fall into three categories: double-digit, single-digit, and below one. The double-digit firms, whose price/sales ratios exceed 10, are all B2B or auction and include Commerce One, Ariba, FreeMarkets, and eBay. The single-digit firms are potential long-term survivors such as Amazon, E-Trade, and VerticalNet. The sub-one firm is Priceline, whose long-term survival is in the most doubt.

These differences provide quantitative confirmation of a trend that is supported by anecdotal evidence. Investors approach Internet stocks in general as if they were rapidly passing fads that have increasingly short

TABLE 7-7. STOCK PRICE AND REVENUE GROWTH PERCENT CHANGE, 2000, AND VALUATION RATIOS FOR SELECTED E-COMMERCE FIRMS, JANUARY 12, 2001

Company	Stock % Change	Revenue Growth	Price/Sales	(Price/Sales) /Rev. Growth	Price/Book
Commerce One	−74%	1,157%	20.92	1.8	2.30
Ariba	−40	515	38.73	7.5	3.10
Sportsline	−89	88	1.53	1.7	1.70
VerticalNet	−92	1,346	3.14	0.2	0.64
E-Trade	−72	232	1.67	0.7	1.98
eBay	−47	97	29.67	30.6	11.31
Amazon.com	−80	86	2.46	2.9	NM
FreeMarkets	−94	338	12.58	3.7	1.79
Priceline	−97	222	0.37	0.2	NM
MEAN	**−76%**	**453%**	**16.08**	**5.5**	**3.26**

Source: MSN MoneyCentral

half-lives. B2B is much hotter than B2C, and as a result, the valuations of all the companies in each group are significantly different. The most likely outcome at this stage appears to be a decline in the valuations of B2B companies as investors continue to grow disenchanted with the category. Furthermore, it is likely that this faddism is likely to continue until firms begin to perform on financial statement categories that can be subject to traditional valuation methods such as price/earnings ratios. Simply put, until firms can generate substantial profits, they are unlikely to escape the rapidly changing whims of investors who flit among Internet investment fads.

The market-valuation metrics also suggest possible mispricing among some of the publicly traded e-commerce firms within these three categories. For example, dividing the revenue growth rate into the price/sales ratio reveals a valuation metric suggesting either that Commerce One is undervalued relative to Ariba or that Ariba is overvalued relative to Commerce One. This same valuation metric raises similar questions regarding eBay and Priceline. In this case, it seems to make sense that eBay would be more highly valued than Priceline, if for no other reason than that eBay has been able to generate profits. However, if Priceline is able to generate a profit, this valuation disparity should narrow.

While these valuation metrics cannot provide us with an objective point estimate of the value of these e-commerce firms, they do generate some useful insights into the different valuations applied to different e-commerce sectors. These metrics also raise questions about different valuations within specific e-commerce sectors that warrant further analysis. In some cases, these valuation metrics may suggest investment opportunities to take advantage of market mispricing. In other cases, the market prices may be appropriate.

The Internet Investment Dashboard analysis suggests that e-commerce is not an attractive place for long-term investors. In Figure 7-10, half of the indicators on the dashboard are dotted, while half are hatched. It is possible that some of these factors will improve over time, but it currently appears that the only attractive investments in e-commerce companies will be a small number of exceptions rather than the average company in the group. As a result, my judgment is that the investment negatives for the average e-commerce company in this sector outweigh the positives.

Our indicators suggest that investors should approach the e-commerce

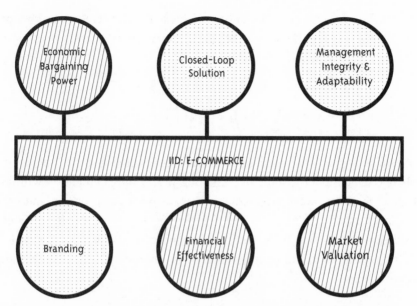

Figure 7-10. Internet Investment Dashboard Assessment of E-Commerce Segment
Key: Dotted = mixed performance and prospects; hatched = poor performance and prospects.

sector with caution. The industry is large and growing, but most companies are losing a significant amount of money, and it is unlikely that most of these firms will be able to charge a price high enough to cover their costs and earn a profit. Despite this lack of pricing power, many e-commerce firms are offering customers very valuable services. As firms run out of money, consolidation is likely. As a consequence, investors who can identify the firms likely to be effective consolidators may be able to profit from their investments in e-commerce. There are also significant variations among e-commerce firms in their management integrity and adaptability and in their branding. Financial effectiveness and relative market valuations, however, appear consistently out of whack.

Nevertheless, there could emerge opportunities for investors who can make astute choices regarding the small number of likely winners in e-commerce. To identify long-term investment opportunities and avoid long-term value destroyers, investors must conduct careful analyses before committing capital to e-commerce. For example, B2B is not likely to support all the firms that have entered the industry in recent years. It may be worthwhile for investors to identify the firms likely to emerge as winners and invest in those. Potential candidates for eventual market dominance may include Ariba and Commerce One.

PRINCIPLES FOR INVESTING IN E-COMMERCE STOCKS

Needless to say, the analysis we have conducted here is subject to change. Undoubtedly, some of the strong-performing companies will falter, some of the weak ones could generate positive surprises, and new competitors could emerge to tilt the playing field in a totally different direction. As a result, investors in e-commerce stocks need some principles to guide them through the analytical process as investment conditions evolve.

E-Commerce Investment Principle 1. Monitor developments in business models that can enable firms to charge prices high enough to earn a profit. As we have noted repeatedly throughout this chapter, the average e-commerce firm is losing lots of money. Until there is significant consolidation among a smaller number of large e-commerce firms that enjoy real economic bargaining power with customers, it is not likely that prices will rise enough to enable e-commerce firms to earn a profit. If this consolidation takes place and firms then raise prices, the profits could be enormous. Another potential scenario for investors to monitor is a significant lowering in the costs of operating an e-commerce business, making it possible for firms to earn profits at existing price levels. While such a scenario is difficult to imagine at the moment, substantial consolidation in the industry could enable surviving firms to maintain their customer bases with much lower levels of spending on sales, marketing, and advertising.

E-Commerce Investment Principle 2. Analyze the evolution of e-commerce firms' strategies to assess which will be able to generate cash flows from their various services. Investors should recognize that many e-commerce firms offer customers closed-loop solutions. The challenge that firms have faced is that they have not been able to get customers to pay for all the individual elements of these closed-loop solutions. Investors will profit from tracking which firms are introducing new services that customers are willing to pay for. Investors may also benefit from monitoring how firms are paring back specific elements of their solutions that are expensive to operate and for which customers are not willing to pay. Simply put, firms that can get customers to pay for their services and that cut back on services that are not worth paying for are likely to do better than those that do not make such distinctions.

E-Commerce Investment Principle 3. Avoid investing in firms where there are questions about management integrity and adaptability. Our analysis indicated questions about management integrity and adaptability in a few firms. This analysis is most useful for investors as a sell signal. If a firm with a track record of avoiding shareholder lawsuits and generating consistent profit growth suddenly deviates from this path, there is a good chance that investors should consider selling the stock. In many cases, such deviations from the path of integrity and adaptability could be a signal that management is no longer up to the job. When these deviations occur, it is likely that management will try to gloss over the problems. Investors may not want to stay along for the ride.

E-Commerce Investment Principle 4. Use compelling brand families as a signal of value in conjunction with other indicators. Prior to the April 2000 crash of e-commerce and other Internet stocks, investors relied rather heavily on brand family in determining which firms to invest in. Such brand families retain their strong signaling value. Increasingly, investors will be able to look beyond such signaling to assess the extent to which the brands can be monetized. Simply put, if the brands help e-commerce firms to generate market-leading revenues, and if these revenues lead to substantial profits, then the compelling brand families are likely to be useful signals of investment value. Companies that are unable to translate their brand families into positive cash flow will not experience rapid stock price appreciation.

E-Commerce Investment Principle 5. Use financial-effectiveness measures to distinguish between the quick and the dead. While investors seem to have tolerated the absence of financial discipline among e-commerce firms up until April 2000, this patience no longer exists. Firms that cannot fund their survival will be liquidated or acquired. Firms that are able to sustain their growth through internally generated funds will predominate.

Investors must conduct a burn-rate analysis to make this distinction. In some cases, firms that can survive until the next capital market window opening are likely to survive even if they are not profitable. Firms that fall into this category are undoubtedly riskier for investors, since the timing of that next window opening is fundamentally unknowable.

E-Commerce Investment Principle 6. Monitor relative valuations among e-commerce sectors and within e-commerce sectors to identify potentially under- and overvalued companies. Relative valuations between e-commerce sectors such as B2B, B2C, and online auctions are likely to fluctuate over time, as we noted earlier. Relative valuations among companies in the same e-commerce sector are also likely to help investors identify specific e-commerce firms whose shares are under- and overvalued. Investors should monitor price-sales, price-sales/sales growth rate, and price-book measures firms within an e-commerce sector. Investors should analyze differences in valuations among firms in sectors such as B2B to assess whether the different valuations are consistent with the underlying fundamentals—as illustrated in their relative performance on the other five Internet Investment Indicators.

CONCLUSION

E-Commerce is unlikely to be an attractive area for investors. Because the industry lacks economic leverage, financial effectiveness, and reasonable market valuations, it is likely that only a comparatively few firms are likely to survive long-term. If an investor can identify the long-term survivors that have a reasonable chance of charging prices that exceed their costs, then there is a good chance of earning attractive investment returns. The foregoing analysis suggests that achieving this investment objective could be very difficult.

Chapter 8

Virtual Doors Slamming Shut:
Web Portal
Stock Performance Drivers

THE KEY INVESTMENT THEME among Web portals is consolidation. There remain one, maybe two Web portals that still retain evidence of their original business vision. Yahoo remains independent; however, some investors are concerned about its dependence on Web advertising from dot-coms. AOL, through its merger with Time Warner, has transformed itself way beyond its original vision; however, its management team remains intact. The other Web portals that were around several years ago have merged out of existence. Investors looking for opportunities in Web portals must recognize the diminishing power of this business model as it was originally conceived.

Web portals are starting points for launching an individual's encounter with the Web. These portals started as search engines and have evolved into media companies and places to transact e-commerce. While Web portals initially sought to accumulate visitors in order to justify high Web advertising rates, they intend to gain an increasing amount of their revenues from e-commerce. This commerce originates from their visitors, is structured through the information that visitors access on the portal, generates a click-through from the portal to a vendor's site, and ultimately leads to a transaction—a piece of which flows through to the Web portals' top line.

During this soon-to-be completed period of consolidation, the stock market performance of the companies being acquired has often exceeded that of the businesses that are likely to be left standing. For example, Lycos saw its stock price increase 56%, from $43 in September 1999 to $67 in September 2000, as a result of the stock market's reaction to its acquisition by Spain's Terra Networks. While Lycos's acquisition by Terra represents some redemption for Lycos CEO Bob Davis, whose controversial merger proposal with Barry Diller's USA Networks fell apart, it remains unclear whether the combination of Lycos and Terra Networks will yield a profitable and rapidly growing business.

By contrast, Yahoo, which is widely perceived as the leading portal, has seen its stock price rise a relatively paltry 36%, from $78 in September 1999 to $106 in September 2000. Yahoo's stock price performance during this year actually represents a 144% decline from its 52-week high of $250 in December 1999. Yahoo has been punished by the cumulative impact of all the negative news that followed the April 2000 Internet stock crash. Many of the dot-coms whose advertising deals constituted a significant portion of Yahoo's revenues ended up in precarious financial shape, jeopardizing Yahoo's anticipated revenue stream. Furthermore, investors began to question the viability of Internet advertising in general, casting doubt on a major source of Yahoo's future revenue stream. Whether investors' concerns prove prescient or overly pessimistic remains to be seen.

For investors, Web portals require some distinct analytical approaches. For example, it is important for investors to understand the relative number of monthly visitors to a Web portal and the Web portals' e-commerce partners. There are three important trends for investors in Web portal companies. First, the Web portal industry has consolidated and is likely to continue to consolidate. Second, the winners are emerging as the only firms left standing—Yahoo is by far the largest portal, and the smaller ones such as Goto.com have limited prospects unless they receive massive amounts of capital from external sources. Third, remaining Web portals are finding that their survival depends on a mixture of revenue sources—increasingly including e-commerce. The effect of this evolution is that Web portals are beginning to resemble e-commerce companies in terms of their investment potential.

This chapter provides investors with the tools needed to analyze these trends and to assess which firms are likely to offer the greatest investment potential.

WEB PORTAL STOCK PERFORMANCE

We begin with an examination of the stock price performance of the Web portal segment. We noted in Chapter 3 that the stock market performance of a cross section of Web portal firms was 167% in 1999. This average performance compares favorably to the NASDAQ's 86% increase in 1999 but represents less than half the 339% average stock price increase of the companies in our nine-segment Internet stock index in 1999.

In 1999, one Web portal, Go2Net, rose over eightfold. As Figure 8-1 indicates, Go2Net's performance was the best among the Web portals. As we noted earlier, consolidation has been significant in this sector. Mergers include At Home and Excite, Disney and Infoseek, CMGI and AltaVista, NBCi and Snap, and Lycos and Terra Networks (to form Terra Lycos). Finally, in October 2000, content and directory provider Infospace purchased Go2Net for $477 million in Infospace stock.

Go2Net is one of two companies mentioned in this book (the other is wireless Internet service provider Metricom) whose stock price benefited tremendously from a substantial investment by Microsoft cofounder Paul Allen. Allen acquired 30% of Go2Net in 1999, and the stock exploded after the announcement. The dynamics of this explosion are worth analyzing. Prior to the announcement, Go2Net was one of several small Web portals generating substantial operating losses whose future appeared uncertain. Following Allen's investment, investors assumed that Go2Net's prospects were assured.

A deeper analysis—one that we will develop in this chapter—suggests that this assumption could be devastatingly wrong. For example, many investors think of Allen as a person with virtually unlimited financial resources who is likely to provide unlimited capital to support his investments. In 2000, Allen disproved this theory when he shut down his R&D operation, Interval Research. Interval Research was intended to be a place where scientists could conduct pure research without concerns for commercialization. Allen's decision to shut down the operation may reflect a growing pragmatism on his part. While Allen may have assumed initially that Interval Research's mission would ultimately produce economic returns, the facts did not bear him out, and he shut down the operation.

A similar pragmatism could well influence his thinking about pump-

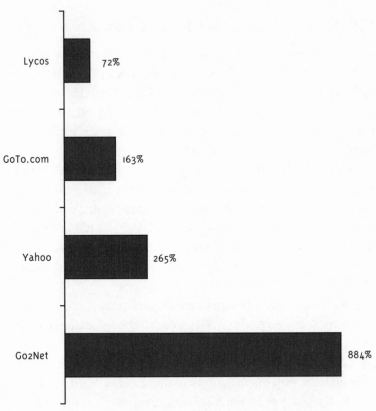

Figure 8-1. Selected Web Portal Firms' Percent Change in
Stock Price, 12/31/98–12/31/99
Source: MSN MoneyCentral

ing additional resources into Go2Net. Based on the decline in Go2Net's stock (as illustrated in Figure 8-2), we should not be overly surprised that Go2Net decided to merge with Infospace. The point to emphasize here is that the driving force behind the 884% rise in Go2Net's stock price may well have been Allen's involvement. This involvement may not justify the stock price increase. As Figure 8-2 indicates, in 2000, investors punished Go2Net and its peers along with most of the other Internet stocks.

The market doled out punishment to Web portals with some degree of consideration for their different states of corporate development. GoTo.com, for example, lost 88% of its value after posting huge operating losses. By the end of 2000 the market had exacted an 81% cut in Terra Lycos's mar-

Figure 8-2. Selected Web Portal Firms' Percent Change in
Stock Price, 12/31/99–5/30/00
Source: MSN MoneyCentral
* Through May 2000

ket value. The market also saw that it had overextended its enthusiasm for Go2Net by chopping away half its value in the spring of 2000. Finally, investors gave Yahoo an 86% haircut during the same period.

As investors contemplate the fate of the Web portal stocks, several factors are most likely to be significant. Relative market share (as measured by the number of average monthly visitors) is likely to remain important for the next several quarters because this number determines the volume of advertising revenues. A more important measure may be the average rate of advertising revenues per visitor. Web portals that offer advertisers the ability to target their advertising to specific groups of customers that are more likely to purchase that advertisers' product should be able to charge higher advertising rates.

Nevertheless, with only 0.6% of Web banner advertisers actually clicking through to the advertisers' site, the rate of return on investment in Web advertising may appear low. This low rate of return makes it increasingly important for investors to analyze how effectively Web portals are broadening their sources of revenue to e-commerce and other nonadvertising sources. Finally, investors are taking a much harder look at profitability, so that Web portals that exhibit greater financial discipline are likely to be long-term survivors and thus to warrant greater consideration from investors.

INTERNET INVESTMENT DASHBOARD ANALYSIS OF WEB PORTALS

While the foregoing discussion of the relative stock price performance of Web portals is suggestive, it does not offer a fully satisfactory explanation. To obtain such an explanation, we must apply the Internet Investment Dashboard (IID) analysis to the Web portal segment.

As we shall see, the IID analysis suggests that the Web portal segment is not likely to generate attractive investor returns in the future. The Web portal industry clearly has virtually no economic bargaining power, as illustrated by its negative margins. Many participants offer customers a closed-loop solution, but they are able to gain a significant share of an unprofitable business. Several industry participants have demonstrated their ability to adapt effectively to change and have created a compelling brand family, but it is not yet clear how these skills translate into profitability. Furthermore, most participants have managed their financial resources based on the assumption that investors would be willing to finance their operating losses. As investor sentiment suddenly shifted in April 2000, bankruptcies and shotgun mergers resulted. This has left a few leaders standing—albeit still struggling to conceive and exploit strategies for profitable operation.

Economic Bargaining Power

As noted earlier, Web portals are targeting three distinct markets—Web advertising, online auctions, and B2C e-commerce. While individually these markets do not offer Web portals much economic leverage, the combined impact of the three may be sufficient to generate attractive profit margins for Web portals. As Figure 8-3 illustrates, U.S. Internet advertising revenues were a modest $2.8 billion in 1999. This amount is forecast to increase almost 10-fold by 2004. Whether this rapid rate of growth can be achieved remains unknown—however, unless Web portals can develop ways to increase the 0.6% click-through rate, it is possible that advertisers will not increase their outlays for Internet advertising so rapidly. Two thousand's dot-com collapse and concerns about a possible recession in 2001 also put the growth of advertising at risk. If Web portals can develop effective methods for matching advertisers with audience segments to increase the rate at which visitors are converted to buyers, then there is room for profit margins in the advertising revenue stream. To date, the extent to which such methods have led to high advertising operating profits has been muted.

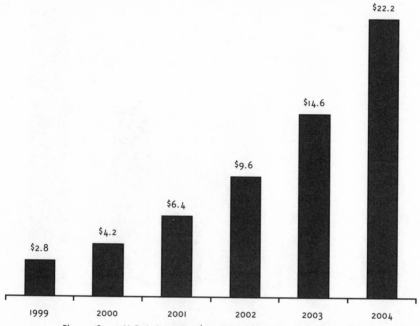

Figure 8-3. U.S. Internet Advertising Revenues, 1999–2004,
in Billions of Dollars
Source: DoubleClick 10K, March 2000

In Chapter 7, we noted that Yahoo was attempting to compete with eBay in online auctions. As illustrated in Figure 8-4, the value of goods sold in online auctions is anticipated to reach $52.6 billion by 2002, growing at a compound annual rate of 56%. As we noted earlier, eBay has clearly established itself as the leader in this segment. Unless eBay slips up in its operations, it is unlikely that Yahoo and other Web portals will be able to generate significant revenues from online auctions. Should Web portals develop effective methods for increasing the level of auction activity on their sites, this additional auction revenue could contribute meaningfully to Web portals' profits.

Web portals' most likely source of revenues to supplement advertising is B2C e-commerce. As Figure 8-5 illustrates, B2C is expected to generate revenues of $145 billion in 2003, growing at a 42% annual rate. But because of competition—since there are hundreds of publicly traded companies whose business models depend heavily on B2C e-commerce—it is not likely that Web portals will enjoy much pricing leverage with con-

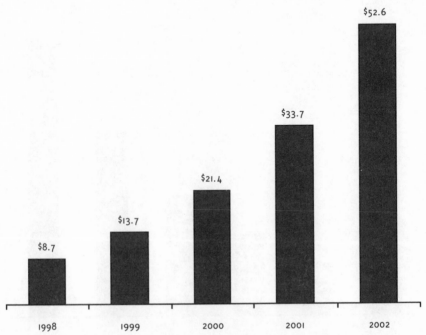

Figure 8-4. Value of Goods Sold in Online Auctions, 1998–2002, in Billions of Dollars

Source: Forrester Research

sumers. However, Web portals will be able to enjoy some profitability from their e-commerce business as a result of the special relationship they have enjoyed with B2C e-commerce companies. Many of these companies have fewer visitors than the Web portals. As a result, these e-commerce companies are willing to pay the portals for click-throughs that originate at the portals and terminate with a transaction at the e-commerce firms' Web sites. The incremental cost to Web portals of these click-throughs is negligible. The incremental revenues for the Web portals are likely to range between 5% and 15% of the amount of the purchase price paid by the consumer. As a result, the potential profitability to the Web portals is very high.

The problem with these transactions lies in the relatively weak bargaining position of many of the B2C e-commerce firms themselves. Following the capital markets' abandonment of unprofitable B2C e-commerce firms in 2000, there is a real risk that the B2C e-commerce firms will run out of money and hence not be able to pay the Web portals the cash to which they are contractually entitled. As a result, Web portals may need to become much more discriminating about the creditworthiness of the e-commerce firms with which they affiliate. In addition, Web portals may evaluate the incremental costs and benefits of backwards-integrating into e-commerce fulfillment activities. Simply put, if there are not enough e-commerce affiliates to generate e-commerce revenues for Web portals, the Web portals may assess whether they can profitably perform these e-commerce activities themselves.

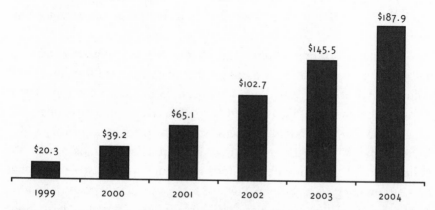

Figure 8-5. Total U.S. B2C E-Commerce Revenues, 1999–2004, in Billions of Dollars
Source: Forrester Research

Web portals have a similarly mixed profitability track record. As Table 8-1 illustrates, the average net margin for the nine Web portal companies sampled in this chapter is −77%. This means that for every dollar that these companies generated in revenue, they lost 77 cents. This statistic suggests the lack of economic bargaining power of the average Web portal company.

TABLE 8-1. MARGINS FOR THREE WEB PORTAL FIRMS, 2000

Company	Stock % Change	Gross Margin	Operating Margin	Net Margin	2-Year Trend
Yahoo	−86	90%	41%	24%	+
GoTo.com	−88	95	−95	−95	+
Terra Lycos	−81	24	−231	−161	−
MEAN	**−85%**	**70%**	**−95%**	**−77%**	**0**

Source: MSN MoneyCentral

Note: Two-year trends represent the extent to which net margins have changed in the most recent two years. + indicates a small improvement, 0 indicates no change, − indicates small deterioration.

The statistic also masks a split in the profit margins of the specific firms sampled. There are clearly two groups of Web portals from the perspective of profit margins. The leading independent Web portal, Yahoo, has actually managed to generate an attractive net margin of 24%. The upstart portal GoTo.com has lost substantial amounts of money in its quest to reach a minimum efficient scale. While the exact point of minimum efficient scale is unclear, this profitability analysis suggests that Yahoo has clearly exceeded that point; Terra Lycos has concluded that it can only reach this point through a merger, and its recent profit performance has been poor.

The analysis of profit margins also highlights an important similarity and an important difference among some Web portals. The important point of sameness for Yahoo and GoTo.com is the gross margins of the firms, which hover at a very high 90% to 95%. These high gross margins suggest that the Web portal business is one that should scale very well. More specifically, if a firm can attract the capital and people needed to build up an infrastructure that supports minimum efficient scale, then the cost of providing each additional unit of service to consumers is quite low and is therefore likely to be very profitable for Web portals. These high gross margins contrast dramatically with those of the B2C e-commerce firms that we examined in Chapter 7 (for example, Amazon.com's gross margin is 26%).

The profit margin analysis also highlights an important difference among the firms. The bulk of the costs associated with building minimum efficient scale are in sales, marketing, and advertising. Simply put, the hard part about getting big is not in the operational infrastructure; rather, it is in convincing millions of consumers to visit the site with sufficient regularity to attract advertisers. Through a combination of partnerships, acquisitions, and marketing Yahoo has managed to build up this base of regular visitors to the point where there is room for profits. These better-established portals are able to use their relatively large number of visitors to persuade advertisers that their money will be well spent on these sites. As a consequence, they do not need to spend as much on sales, marketing, and advertising to add new visitors. By contrast, GoTo.com is spending huge amounts of its revenues on such traffic-building investments.

This analysis suggests an important set of forces likely to drive consolidation in the Web portal sector. Below a minimum efficient scale, a vicious cycle threatens to drive Web portals into the hands of deeper-pocketed partners. Above that minimum efficient scale, a virtuous cycle is likely to drive ever-accelerating revenue growth

The vicious cycle works as follows. The Web portal starts well behind industry leaders; thus it must spend money to generate more appealing content that will attract consumers. But many consumers are already accustomed to using the leading portals, and as a result they may be reluctant to switch to these new portals or to add the new portals to the list of sites they visit. This reluctance to switch raises costs for the Web portals, making it more difficult for them to attract new consumers. The lower number of additional consumers makes it more difficult to attract advertisers, which may have already struck deals with leading portals. The result is lower revenues for the lagging portals, which must invest increasingly high proportions of their revenues to attract new consumers. And so the vicious cycle forces lagging Web portals to spend more and more money to attract fewer consumers until they run out of money.

The virtuous cycle, by contrast, makes it easier for leading Web portals to attract more consumers and advertisers. When a new consumer signs up to use the Web for the first time, the consumer is likely to ask friends which portal has the best content. The friends are likely to recommend the portal that they already use, which is more than likely to be one of the leading portals. As a result, the leading portals are likely to get a disproportionate share of the new Web users, thus increasing their leadership position. The self-reinforcing cycle of market leadership will appeal to businesses that are seeking to minimize the number of portals on which they advertise—focusing their attention on a smaller number of portals that reach the biggest audiences. The money from advertisers provides additional fuel for the leading portals to use to form partnerships, make acquisitions, and advertise and promote their portals in order to attract more visitors. As long as advertisers spend cash to promote themselves with portals, this virtuous cycle can expand Web portals' value.

The implications for investors are clear. Investors should seek out firms that are benefiting from the virtuous cycle and shun those firms that are spiraling downward as a result of the vicious cycle. At first blush, some value-oriented investors may think that an investment in a beaten-down Web portal may provide a profit opportunity as an acquisition. As a practical matter, however, the chances are that a Web portal spiraling downward would be acquired by a vulture acquirer only after the firm reached the verge of liquidation, where the firm's assets could be picked up at the lowest possible price. Investors should recognize that the firms benefiting from the virtuous cycle have a better chance of appreciating in value over the long run.

At this point in their evolution, Web portals' economic bargaining power is mixed at best. In advertising, which represents the bulk of their revenues, Web portals have low to moderate bargaining power because the click-through rates are low. The larger Web portals are able to offset this low click-through rate by citing their relative market share. Obviously, this factor enhances the bargaining leverage of Web portals in proportion to their relative size. While auctions are likely to be small as a proportion of revenues, they are likely to generate fairly high profit margins because the incremental cost of servicing each auction transaction is low. Finally, the profit potential of e-commerce is likely to be high as long as Web portals' affiliates can pay the fees to which they are contractually obligated.

Closed-Loop Solution

One thing that many Web portal firms have done successfully is to offer customers a closed-loop solution. For example, a visitor to Yahoo can search the Web, track stocks, check on the latest news, and launch a variety of e-commerce transactions, including stock trades. This enhanced value proposition is a prescription for rapid revenue growth.

One way to assess the relative effectiveness of the strategies of firms in the Web portal market is to analyze the firms' relative size, revenue growth, and market capitalization. This analysis can help us gain insights into which firms are generating the greatest product market momentum and the extent to which that momentum has translated into greater stock price appreciation.

The correlation between momentum in the product markets and in the capital markets is not really borne out by the evidence. As Table 8-2 indicates, it appears that a Web portal's rate of shareholder value destruction is not correlated with its rate of revenue growth. For example, the fastest-growing firm in terms of revenue (at a 938% annual rate) was Terra Lycos, whose stock price declined at the third-fastest rate of the firms in our sample. However, the firm with the slowest revenue growth, Yahoo (at an 88% rate), experienced the second slowest rate of stock price decline.

TABLE 8-2. REVENUE MOMENTUM OF THREE WEB PORTAL FIRMS, JANUARY 12, 2001, IN MILLIONS OF DOLLARS

Company	Stock % Change	12-Months Revenue	Revenue Growth	Market Capitalization
Yahoo	−86	1,100	88	14,400
GoTo.com	−88	77	367	535
Terra Lycos	−81	119	938	6,510

Source: MSN MoneyCentral

What is most striking about this table is the relative market capitalizations of the Web portal firms. Yahoo, which is almost 10 times as large in revenues as its nearest competitor, Terra Lycos, has a market capitalization that is twice that of Terra Lycos. This analysis suggests that the stock market may be undervaluing Yahoo and overvaluing Terra Lycos.

In order to gain meaningful insights that can explain how the differences in strategies of the firms translate into different levels of return in the stock market, it is important to analyze the different strategies of the Web portal firms. Table 8-3 summarizes the key strategic differences among the three firms profiled in this chapter.

TABLE 8-3. STRATEGIC DIFFERENCES AMONG FOUR WEB PORTAL FIRMS

Company	Competitive Strengths/Weaknesses
Yahoo	*Strengths:* 120 million visitors worldwide per month, 3,800 advertisers, global presence, excellent management, profitable, backed by Softbank
	Weaknesses: Lack of subscription revenue creates competitive disadvantage relative to AOL
GoTo.com	*Strengths:* 29,000 advertisers bid for space on search engine; broadening through acquisitions into auction and shopping
	Weaknesses: Weak financial performance, small
Terra Lycos	*Strengths:* 91 million monthly visitors; manages multiple Web properties; profitable; Terra Networks deal strengthens global presence
	Weaknesses: Not a market leader

Source: Company Reports, Peter S. Cohan & Associates Analysis

This analysis of Web portal business strategies suggests that there is some correlation between the Web portal firm's relative market capitalization and the robustness of its competitive strategy. For example, Yahoo, with 120 million visitors per month worldwide, has a market capitalization twice that of the number-two Web portal, Terra Lycos, which has 91 million monthly visitors.

GoTo.com has a strategy that is interesting in its difference from competitors' strategies—even though it does not generate sufficient revenues to make GoTo.com a significant player. More specifically, GoTo.com is more targeted toward advertisers by offering them a chance to bid for rank in topic searches instead of asking them to pay for banner advertisements or site sponsorship. GoTo.com's approach works as follows: Five online stock brokers each bid for their name to appear first in response to a GoTo.com visitor search on "online brokers." While this approach is intended to increase revenues for GoTo.com and improve the value of "advertising" from the perspective of sponsors—because of the presumed higher level of consumer motivation to buy from the sponsor—its ultimate success depends on attracting a larger number of consumers. Regrettably for GoTo.com shareholders, it is not clear that the strategy is creating superior value for consumers. As a result, there is a chance that GoTo.com will not generate the volume it needs to reach minimum efficient scale.

For investors in Web portals, it is important to look more closely at which firms have the highest relative market capitalization. The big firms are likely to survive because they have access to a broader range of options for capital raising and for strategic diversification. More specifically, the firms with the larger market capitalizations will be able to raise additional capital by selling shares to investors, through mergers with cash-rich partners, or through infusions of private capital. Furthermore, firms with larger market capitalizations still have the currency—in the form of relatively expensive shares—needed to acquire companies that can give these larger players access to more customers and more services around the world.

The foregoing discussion is important because investors in Web portal companies have assigned the highest market capitalizations to the firms that do the best job of offering closed-loop solutions. These closed-loop solutions must, by the very nature of Web portals, focus on consumers and advertisers. For consumers, the closed-loop solution means access to a broad range of services that are most useful to them; therefore, they will stay at the portal for long periods of time.

For advertisers, the closed-loop solution means helping to target their advertising and promotional activity on the site to segments of consumers with the highest propensity to buy the advertisers' products or services. The more effectively the Web portal serves consumers, the more likely that portal is to attract increased advertiser attention. The effect of the virtuous cycle we described earlier is to create measurable increases in Web portal visitors—and subsequent increases in relative market capitalization. For investors, this relative market capitalization is the best predictor of long-term survival. Similarly, the best predictor of a firm dropping out of the long-term race is whether the firm's size is below the minimum efficient scale and is thus caught in a vicious cycle—not the short-term impact of specific investments, such as Paul Allen's deal with Go2Net.

Management Integrity and Adaptability

Management integrity and adaptability are important in the Web portal segment. Consumers and advertisers are particularly dependent on Web portal managers' ability to meet their commitments and to anticipate and

adapt to changing technology, customer demands, and competitor strategies. Given the rate at which various portals can replicate innovative fea-
tures that consumers and advertisers value, it is particularly important for
investors to be able to assess how well management adapts to changes in
the competitive environment.

An analysis of legal proceedings and accounting policies for the four
Web portal firms suggests that while many of the firms are involved in low
levels of litigation, there is no firm that stands out as unusually challenged
in its management integrity. As Table 8-4 indicates, the Web portal firms
employ conservative accounting techniques which—if applied consistently—suggest that investors will not need to worry about overstated Web
portal revenues.

TABLE 8-4. REVENUE ACCOUNTING POLICY AND SHAREHOLDER LAWSUITS FOR SELECTED WEB PORTAL COMPANIES

Company	Revenue Accounting Policy	Legal Proceedings
Yahoo	Advertising revenues are recognized over the period in which the advertisements are displayed. Deferred revenues result from billings in excess of recognized revenue relating to contracts.	Yahoo has been subject to claims of alleged infringement of trademarks, copyrights and other intellectual property rights, claims arising in connection with its e-mail, message boards, auction sites, shopping services, and other communications and community features, such as claims alleging defamation or invasion of privacy. Yahoo has also been advised that the FTC is conducting an inquiry into certain of Yahoo's consumer information practices.
GoTo.com	Search listing revenue is recognized when earned based on click-through activity to the extent that the advertiser has deposited sufficient funds with GoTo.com or collection is probable.	GoTo.com was involved in litigation against Walt Disney and Infoseek alleging infringement of the GoTo.com logo. On May 25, 2000, Disney paid $21.5 million to GoTo

Company	Revenue Accounting Policy	Legal Proceedings
		and discontinued use of the GoTo.com logo.
Terra Lycos	Lycos' advertising revenues are derived principally from short-term advertising contracts in which Terra Lycos guarantees a number of impressions for a fixed fee or on a per impression basis with an established minimum fee. Revenues from advertising are recognized as the services are performed.	Terra Lycos is involved in lawsuits alleging misrepresentations and omissions relating to a transaction with USA Networks in 1999, which was ultimately canceled.

Source: Company 10Ks

The lawsuits filed against these Web portals do not suggest significant ethical problems. An analysis of the different lawsuits does suggest areas that might emerge as concerns for investors in the future. For example, Yahoo appears to have left itself open to regulators concerned about violation of individual privacy. To the extent that Yahoo's revenues depend on selling information about consumers' Web activity to advertisers, the litigation alluded to in Table 8-4 could cost Yahoo some of its future revenue growth.

GoTo.com's settlement of its litigation with Disney provided cash and relieved the pall over the stock caused by Disney's having had much deeper pockets than GoTo.com and therefore being in a stronger position to sustain the litigation in an effort to distract GoTo.com's management from growing revenues. Unfortunately for GoTo.com shareholders, the settlement did not keep GoTo.com's stock price from losing half its value between May and November 2000.

Finally, Terra Lycos's litigation related to the collapsed USA Networks deal reminds investors of Terra Lycos's apparent overeagerness to conclude a merger that increases Lycos's scale and preserves a job for the company's CEO, Bob Davis. In light of the significant drop in the stock price of Terra Networks, the new acquiring firm, and the meaningful management challenges associated with building the combined company, it remains unclear how successful the merger with Terra Networks will be.

In light of the rapid rate at which the competitive environment changes, the ability of managers to adapt to this change effectively is a crucial factor in determining the long-term value of an Internet company. To analyze this factor, we can examine the variance of the percentage change in quarterly earnings. If a Web portal firm is able to increase profits fairly consistently in each quarter, its management is good at adapting to change. Conversely, if a firm's quarterly performance jumps around significantly, its management team is probably less effective at adapting to change. The important point here is that the ability to adapt effectively to change in a rapidly changing environment can be measured by management's ability to produce fairly consistent financial results.

In the case of Web portal firms, this measure appears useful, not so much as a predictor of relative stock price appreciation but as an indicator of long-term viability. As Figure 8-6 indicates, the firm with the most consistent earnings growth is Yahoo and the one with the most inconsistent performance is Terra Lycos. As noted above, Terra Lycos has a track record of stretching for the big deal that will transform its lagging position into one of world dominance. This stretching has proven to be wishful thinking in the past and has contributed to inconsistent earnings performance. By contrast, Yahoo's 20% variance in quarterly earnings change reflects its ability to control its destiny and reflect this control in relatively predictable results.

As suggested earlier, the concept of the Web portal has changed very dramatically over the last several years. For example, Web portals were originally search engines with catalogs of new Web sites. Through acquisitions and partnerships, Web portals added capabilities such as free e-mail, shopping services, chat, home pages, and personalization. All industry participants adapted quickly to these changes. As noted earlier, some participants (e.g., Excite, Infoseek, and Snap) decided that their best end-game strategy was to seek out an acquirer. Others maintained their independence and expanded through acquisition. The measure of earnings consistency summarized in Figure 8-6 is a good shorthand measure for investors to identify the firms that have been effective at adapting to this change and those that have been less effective.

Management integrity and adaptability are critical factors in differentiating winning Internet companies from their peers. The evidence collected here suggests that Web portals score well on accounting policy but have a more varied record in the handling of lawsuits. The litigation suggests future challenges without necessarily impugning the integrity of

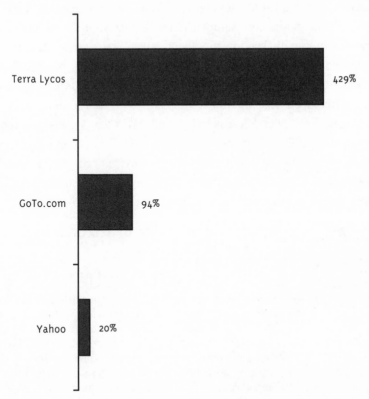

Figure 8-6. Variance in Quarterly Earnings' Percent Change for
Selected Web Portal Firms, 3/99–3/00
Source: MSN MoneyCentral, Peter S. Cohan & Associates Analysis

management. These challenges range from potential concerns about Yahoo's future ability to trade on information about its visitors to uncertainties regarding the ability of Lycos to conclude its deal with Terra Networks. The earnings-consistency analysis suggests that Yahoo is by far the most effective Web portal at adapting to change. Yahoo started off in the lead and has sustained that lead.

Brand Family

Web portals can be judged to a large extent on the basis of the Web sites that they have assembled and the nature and quality of their partners and investors. As Table 8-5 indicates, there is significant variation in the quality of Web portals' Web sites and partnerships. Firms seem split between those that have established leadership and those that have followed through partnerships with laggards. For example, Yahoo has assembled a variety of content partnerships with leading providers in each category. Firms such as Lycos have attempted to establish partnerships with many providers in the same content category, such as stock price information and market analysis. The key difference is that Yahoo does business with the top-tier providers in each category, while Terra Lycos works with firms that are not as strong in their categories. GoTo.com is pursuing an entirely different approach, which includes revenue-sharing arrangements with lesser portals. Terra Lycos has also negotiated e-commerce partnerships with some well-known brand names; however, Terra Lycos's partnership strategy lacks the depth of Yahoo's.

TABLE 8-5. MAJOR SITES/PARTNERS OF SELECTED WEB PORTAL COMPANIES

Company	Major Sites/Partners
Yahoo	Yahoo integrates content into subject-based properties from Reuters New Media, Associated Press, Deutsche Presse Agentur, and Agence France Presse, stock quotes (Reuters), corporate earnings reports (Zacks), audio news (National Public Radio), mutual fund holdings (Morningstar), stock investing commentary (Motley Fool, CBS MarketWatch, The Street.com), sports scores (ESPN SportsTicker), sports commentary *(The Sporting News)*, weather information (Weathernews, Inc. and the Weather Channel), and entertainment industry gossip (E! Online). Among new properties Yahoo launched in 1999 are Yahoo Health, Yahoo Pets, Yahoo Politics, and Yahoo Entertainment; Yahoo offers its members the opportunity to purchase automotive services (Autoweb.com), books (Amazon.com), brokerage services (E*Trade, TD Waterhouse, and National Discount Brokers), flowers (FTD, Gerald Stevens), health care (Health Network, Healtheon/WebMD), mortgages (E-Loan), music (ARTISTdirect), traditional communications services (AT&T), and wedding-related products and services (WeddingNetwork, WeddingChannel.com); Yahoo Shopping includes 7,500 stores, including Brooks Brothers, Coach, Eddie Bauer, The Gap, Guess?, Nordstrom, Patagonia, Macy's, OfficeMax, Toys R Us, Victoria's Secret, and Zales. Yahoo has relationships with Covad Communications and NorthPoint Communications. Yahoo also partnered with Kmart's BlueLight.com and Spinway.com to deliver free Internet access to Kmart shoppers.

Company	Major Sites/Partners
GoTo.com	Click-through deals with Netscape, Microsoft, EarthLink, MindSpring, Dogpile, OneMain
Terra Lycos	Terra Lycos partners include Coca-Cola, Disney, Dell, The Gap, Intel, Sony, and Visa. Terra Lycos has established electronic commerce and sponsorship relationships with Barnes & Noble, First USA Bank, Fleet Bank, and WebMD. In addition, Lycos has established strategic licensing and technological alliances with Fidelity Investments, IBM, Microsoft, Packard Bell/NEC, RCN, and Viacom

Source: Company Financial Statements (10Ks)

Differences among Web sites and partners seem to correlate with differences in relative market capitalization, not stock price performance. Just as our analysis of management adaptability indicates that Yahoo's effective adaptation translates into a higher market capitalization, so does its outstanding partnership strategy correlate with its leading market capitalization. Terra Lycos's Web sites and partnership strategy are that of a follower.

The quality of a Web portal firm's investors seems to tell us little about the relative stock price performance. As Table 8-6 indicates, with the exception of Lycos (which does not have any major shareholders in management), Web portals' major shareholders are a combination of insiders

TABLE 8-6. MAJOR INVESTORS IN SELECTED WEB PORTAL COMPANIES

Company	Major Investors
Yahoo	Softbank (22.6%), David Filo (8.7%), Jerry Yang (8.4%), Timothy Koogle (1.8%)
GoTo.com	Bill Gross's idealab! (27.3%), idealab! Capital Management (6.2%), Gilder, Gagnon, Howe & Co (5.6%), Moore Capital Management (8.1%), Jeffrey S. Brewer (3.5%), Timothy Draper (3.8%), William Elkus (6.2%), Linda Fayne Levinson (4.1%)
Lycos	CMGI (16.6%), FMR (11.8%).

Source: Company Proxy Statements

and venture capital firms. Yahoo's major venture capital firm investor is Masayoshi Son's Softbank; GoTo.com's major venture capital investor is Bill Gross's idealab; and Lycos's major venture capital investor was David Wetherell's CMGI prior to the merger with Terra Networks.

It is interesting to note that each of these venture firms has a leader around which the media has been enlisted to create somewhat of a cult following. Major media have written stories praising the investment genius of each of the names associated with these venture capital firms. As we will explore in Chapter 10, there may be some interesting ownership relationships between the media that spin the puff pieces and the venture investors themselves. Simply put, these "investment geniuses" may have some inordinate degree of control over the media people who are writing the stories. The implication for investors in Web portals is that they should pierce the magic veil associated with the name of the big investor and examine Web portals' other business fundamentals.

Financial Effectiveness

Prior to the second quarter of 2000, investors did not use financial effectiveness as a means of discriminating among Web portal firms. While there is no correlation between Web portal firms' financial effectiveness and their stock market performance, there is once again a correlation between financial effectiveness and relative market capitalization.

There is a wide variation among three classes of Web portal firms in terms of their sales productivity. As Figure 8-7 suggests, the productivity figures should be analyzed in two groups—those of Web portals that have reached minimum efficient scale and those that have not. Yahoo and Lycos, which have reached minimum efficient scale, have higher sales productivity than GoTo.com, which has not.

Yahoo's sales productivity is significantly higher than that of Terra Lycos, possibly owing to its tighter budgeting processes. GoTo.com's sales productivity is by far the lowest of the Web portals, suggesting how difficult it could be for GoTo.com to reach minimum efficient scale with its current business model.

The correlation between SG&A to sales and relative market capitalization is equally pronounced. As Figure 8-8 (page 210) suggests, the firm with the lowest SG&A-to-sales ratio has the highest market capital-

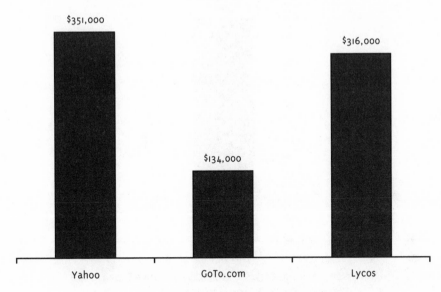

Figure 8-7. Sales per Employee of Three Web Portal Firms, 1999
Source: MSN MoneyCentral

ization. With the exception of Terra Lycos, the most efficient firms are rewarded with the highest market capitalizations. Terra Lycos, while relatively inefficient, is much bigger in terms of revenue than GoTo.com and is thus rewarded with a higher market capitalization despite its higher SG&A-to-sales ratio.

The conclusion from this analysis is that investors have focused on financial effectiveness among Web portal firms in conferring relative market capitalization but not relative stock price performance. Yahoo has the highest sales productivity, the lowest SG&A-to-sales ratio, and the highest market capitalization. GoTo.com has the lowest sales productivity, the highest ratio of SG&A to sales, and the lowest market capitalization. The implications for investors is to focus on Web portals with the greatest financial effectiveness because those firms are most likely to survive in the current climate, where capital is driven more by fear of loss than by fear of missing gains.

Relative Market Valuation

The stock market's approach to valuing Web portal companies suggests that investors recognize the relative merits of Yahoo's business

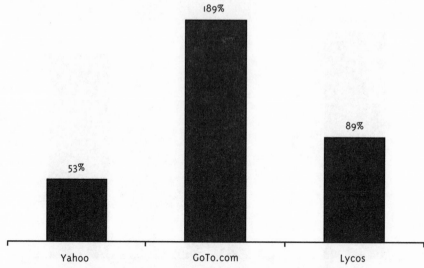

Figure 8-8. Selling, General, and Administrative Expense as a
Percent of Sales for Three Web Portal Firms, 1999
Source: MSN MoneyCentral

model. As Table 8-7 indicates, on the basis of price/sales ratios, at 30
times sales, Terra Lycos is valued more than two times higher than its
nearest peer, Yahoo.

**TABLE 8-7. STOCK PRICE AND REVENUE GROWTH PERCENT CHANGE,
2000, AND VALUATION RATIOS FOR SELECTED WEB PORTAL FIRMS,
JANUARY 12, 2001**

Company	Stock % Change	Revenue Growth	Price/Sales	(Price/Sales) /Rev. Growth	Price/Book
Yahoo	−86	88	14.0	15.9	7.64
GoTo.com	−88	367	6.2	1.7	1.20
Terra Lycos	−81	938	29.8	3.2	3.42
MEAN	**−85%**	**464%**	**16.7**	**6.9**	**4.09**

Source: MSN MoneyCentral

It is interesting to note, however, that on a (price/sales)/revenue
growth basis, Yahoo is almost five times more valuable than Terra Lycos.

This suggests that investors may be struggling with a tradeoff between relative size, where Yahoo has an advantage, and relative revenue growth, where Terra Lycos enjoys an edge. Since Terra Lycos's high growth is attributable to the recent merger, the sustainability of this growth is uncertain.

The Internet Investment Dashboard analysis suggests that Web portals could be an attractive place for long-term investors. In Figure 8-9, all the indicators are dotted. The most critical factor that investors must consider is whether an individual firm has reached the scale and relative market share needed to enjoy economic bargaining power with advertisers and e-commerce partners. The challenge for investors is that firms like Yahoo, which clearly have this bargaining power, may be fully valued by investors and may therefore have limited upside potential. By contrast, firms below the minimum efficient scale may have great difficulty charting a course that enables them to leap over the bar.

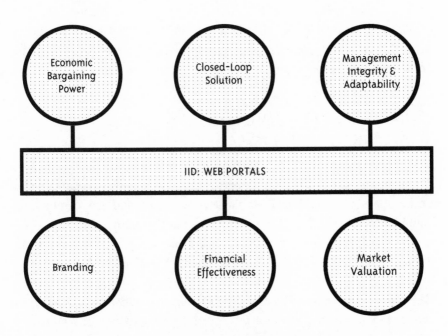

Figure 8-9. Internet Investment Dashboard Assessment of
Web Portal Segment
Key: Dotted = mixed performance and prospects.

The indicators in Figure 8-9 suggest that investors should approach the Web portal sector with caution. The industry is evolving from one dependent solely on advertising to one that derives revenues from a variety of sources, including e-commerce and auctions. Web portals that attract large numbers of visitors can charge premium prices to advertisers and e-commerce partners. Smaller Web portals may lack this pricing power and therefore must spend a higher proportion of their limited resources on sales and marketing. Most Web portals offer closed-loop solutions to customers, and mobility barriers are fairly low. As a result, innovations by one Web portal can generally be replicated rather quickly by competitors. The level of management integrity and adaptability among Web portal firms is generally high, although many firms are involved in litigation that suggests potential concerns. Web portals generally do a good job of building their brand families, although once again the largest firms are able to command partnerships with the biggest-name partners and the smaller firms end up partnering with less-than-top-tier names. Financial effectiveness seems to improve with size: Yahoo has the highest sales productivity and lowest SG&A-to-sales ratio. Finally, relative valuations suggest that firms' market capitalizations and price-to-sales ratios correlate strongly with the relative performance of Web portals on the five other IID indicators. As a result, the upside potential for investing in Web portals may be muted.

PRINCIPLES FOR INVESTING IN WEB PORTAL STOCKS

Needless to say, the analysis we have conducted here is subject to change. Undoubtedly, some of the strong-performing companies will falter, some of the weak ones could generate positive surprises, and new competitors could emerge to tilt the playing field in a totally different direction. As a result, investors in Web portal stocks need some principles to guide them through the analytical process as investment conditions evolve.

Web Portals Investment Principle 1. Monitor consolidation and partnership trends to identify changes in the list of firms that exceed minimum efficient scale. As we have noted throughout this chapter, the concept of minimum efficient scale, while difficult to quantify precisely, plays an important role in determining whether a Web portal can operate profitably.

The most significant investment opportunities may be firms that are likely to chart effective paths to exceeding minimum efficient scale. For example, if GoTo.com could execute a strategy—possibly through mergers or strategic alliances—that would generate much higher numbers of visitors, then GoTo.com could appreciate dramatically in the stock market. As noted earlier, scale confers pricing power on Web portals that are able to provide measurable value to advertisers and e-commerce partners by giving these partners access to huge audiences of consumers. Furthermore, investors should shy away from Web portals if growth in Internet advertising decelerates.

Web Portal Investment Principle 2. Analyze the evolution of Web portal firms' strategies to add new services that result in new revenue streams. Investors should recognize that many Web portal firms offer customers closed-loop solutions. The challenge that firms have faced is that they have not been able to get customers to pay for all the individual elements of these closed-loop solutions. Investors will profit from finding out which firms are introducing new services that customers are willing to pay for. Investors may also benefit from monitoring how firms are paring back the elements of their solutions that are expensive to operate and for which customers are not willing to pay. Simply put, firms that can get customers to pay for their services and that cut back on services customers feel are not worth paying for are likely to do better than firms that do not make such distinctions.

Web Portal Investment Principle 3. Monitor Web portals' management integrity and adaptability. Our analysis indicated that Web portals score generally well on management integrity and adaptability. As Web portals increasingly diversify their sources of revenue, investors should pay close attention to their policies for revenue accounting to be sure that firms are not artificially inflating revenues, particularly in the area of e-commerce. Furthermore, investors should monitor how ongoing litigation is resolved. For example, if Yahoo is forced to stop selling information about its visitors, the change could reduce Yahoo's revenue outlook.

Web Portal Investment Principle 4. Use compelling brand families as a signal of value, but track how these signals are converted into cash. The quality of Web portals' partners tends to vary with relative market share. Simply put, the Web portals that attract the most visitors are able to attract the most well-known advertisers and forge e-commerce partnerships with the biggest retailers. Following the April 2000 crash of Internet stocks, investors became increasingly likely to seek quantitative evidence that these compelling brand families can be converted into accelerated revenue growth. Evidence of such accelerated growth—potentially in the form of a decrease in the ratio of incremental sales and marketing expense per new site visitor—could precede a rise in that Web portals' stock price.

Web Portal Investment Principle 5. Use financial-effectiveness measures to determine whether a firm exceeds minimum efficient scale. Our analysis clearly indicated a split between big Web portals and small ones. The big firms have much higher sales per employee and much lower SG&A-to-sales ratios. Investors should monitor how these indicators change over time to identify whether—or when—specific firms are improving their operations enough to warrant higher stock prices and price-to-sales ratios. Significant improvement over time could provide investors with an early-warning indicator that precedes stock price appreciation.

Web Portal Investment Principle 6. Monitor relative valuations among both big and small Web portals to identify potentially under- and overvalued companies. Big Web portals are likely to have higher valuations than small Web portals. Within the big Web portal sector, the price-to-sales ratios of individual firms should be consistent with the firms' relative performance on the other five IID indicators. If a big Web portal's stock market valuation pulls away from its relative performance on the other indicators, this disparity could indicate a selling opportunity in that particular security. Conversely, if a big Web portal's stock market valuation dips far below its relative performance on the other indicators, this disparity could indicate a buying opportunity.

CONCLUSION

Web portals constitute an area of muted attractiveness for investors. Given the small number of remaining industry participants, it is likely that the winners are fully valued. Furthermore, the challengers have limited upside price-appreciation potential unless they are able to execute strategies that can dramatically increase their relative market share to the point where they exceed minimum efficient scale. To the extent that investors can identify such transformational strategies before they are fully reflected in stock prices, Web portals could be an area of investment opportunity.

Chapter 9

KEEPING THE BARBARIANS
BEHIND THE GATES: WEB SECURITY
STOCK PERFORMANCE DRIVERS

WEB SECURITY STOCKS ARE clearly divided into the acquirers and the pack. Successful acquirers create virtuous cycles that enable their stock prices to rise, while those in the pack are caught in the winners' backwash. As we will see in a moment, acquiring is not always easy to do. If done well, however, acquisitions in the Web security business can accelerate revenue and profit growth, endowing the successful acquirer with a higher stock market capitalization than its peers. This higher market capitalization provides the currency needed to make further acquisitions, which, if properly planned and executed, can lead to even faster revenue and profit growth.

VeriSign is one of the acquirers whose stock price has increased. Between September 1999 and September 2000, VeriSign's stock appreciated 236%, from $50 to $168. This performance, while strong, also represents a 35% decline from VeriSign's 52-week high, following the March 7, 2000, announcement of its intention to acquire Network Solutions for about $21 billion. Prior to the Network Solutions announcement, VeriSign's stock peaked at $258. This deal closed in June 2000, and the full financial impact remains to be seen. Clearly, investors began to doubt VeriSign's vision—by the end of 2000, its stock had declined to $74, a 61% drop during the year. Nevertheless, time will tell whether the strategic benefits of the acquisition offset its costs.

In the judgment of the stock market, Network Associates is a clear-cut example of how not to make acquisitions in the Web security market. Between September 1999 and September 2000, for example, Network Associates' stock appreciated a relatively paltry 11%, from $20 to $23. During that year, Network Associates bounced around from a low of $16 to a high of $37. The key point for investors to understand is that Network Associates, despite its relatively large base of revenues, has been unable to use its acquisitions to accelerate revenue and profit growth. As a result of Network Associates' relatively ineffective job of acquisition planning and integration, its stock has been stuck in the mud for years. Network Associates' stock peaked at $65 in January 1999, then declined to a low of about $12 in April 1999. Network Associates blew its credibility with investors and in 2001 replaced its senior management in an effort to regain that credibility.

In recent years, occasional hacker attacks have received tremendous publicity. For example, in April 2000, the "ILOVEYOU" virus, spread by e-mail, received a tremendous amount of attention from the press. Many users of Microsoft's Outlook Express e-mail software received e-mails with "ILOVEYOU" in the subject line. When recipients opened the file attached to the e-mail, the virus sent a copy of itself to all the e-mail addresses in that recipient's address book. The effect of the virus was to shut down the e-mail functions of organizations around the world as the increase in e-mail volume exceeded capacity.

The long-lasting result of such highly publicized attacks is to fuel demand for products and services that can protect organizations from future attacks. In fact, it is interesting to note that the stock prices of leading Web security firms seem to increase after particularly well-publicized attacks. It would be convenient for investors if they could simply identify leading Web security firms and then buy their shares the day before such a hacker attack was announced. However, such investment opportunities exist only in the world of imagination.

Web security is a combination of technology and services intended to help organizations protect themselves from computer hackers. There are several common ways that hackers attack networks. Hackers can introduce viruses, as we noted earlier, that can seriously weaken network performance. Hackers can break through organizations' network "firewalls" to steal confidential information. Hackers can create false identities that enable them to gain unauthorized access to confidential information.

Hackers can also intercept confidential information during its movement from a source to a destination.

Web security firms provide products and services that are designed to protect against such hacker attacks. Web security firms test a networks' degree of vulnerability to hacker attacks, identify the points of greatest vulnerability, and recommend ways to bolster network security. Firewall software is intended to make sure that unauthorized insiders and outsiders cannot gain access to confidential information. Antivirus software is intended to block viruses from gaining access to networks. Authentication software verifies the identity of a person who sends a message. Encryption software scrambles information so that if it is intercepted, the hacker cannot make sense out of the information.

WEB SECURITY STOCK PERFORMANCE

We begin with an examination of the stock price performance of the Web security segment. We noted in Chapter 3 that the stock market performance of a cross section of Web security firms was 263% in 1999. While this average performance compares favorably to the NASDAQ's 86% increase in 1999, the 263% increase compares quite unfavorably to the 339% average stock price increase of the companies in our nine-segment Internet stock index in 1999. The 2000 performance of this sector, down 26%, makes it one of the least negative of the Internet stock sectors.

One important factor behind the performance of Web security stocks is the mix of winners and losers in the industry. Web security is a relatively fragmented industry that makes room for many public companies, some of which are not particularly successful. While there continues to be consolidation in the industry, there are several firms that survive despite weak financial performance. These weaker firms—such as Secure Computing, V-One, and Network Associates—drag down the performance of the average. Interestingly, it appears that some of these firms are in such weak strategic positions that they are not even attractive acquisitions targets.

Despite the weak firms, the stock market performance of the strong firms is so great that the average performance of the sector was almost three times better than the NASDAQ in 1999 and was one of the few Internet stock sectors that declined less than the NASDAQ's 39% drop in 2000.

As Figure 9-1 indicates, firms like VeriSign and Check Point Software helped raise the overall average performance of the Web security stocks. While the financial performance of these two leaders is very different, they are both very highly regarded in terms of their market position, the quality of their products and services, their partnership strategy, and their management. VeriSign is the leader in trust services: It helps authenticate e-commerce transactions through a combination of technology and security processes that create a tightly protected network. While VeriSign has not been profitable, its revenues have grown very rapidly. Check Point Software is the leader in firewalls and has formed partnerships with many

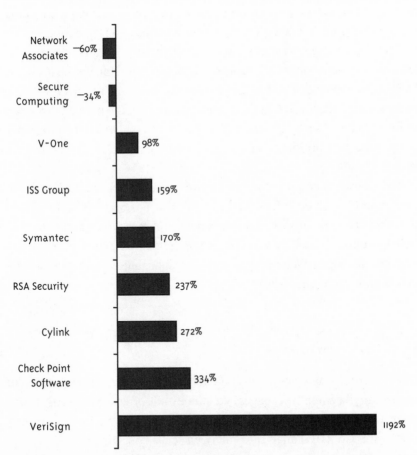

Figure 9-1. Selected Web Security Firms' Percent Change in Stock Price,
12/31/98–12/31/99
Source: MSN MoneyCentral

firms in order to broaden the usefulness of its products. Check Point's profitability and revenue growth have both been exceptionally high.

The effect of the April 2000 crash in Internet stocks had a somewhat unusual effect on the Web security firms' stock prices. The market did not uniformly destroy the value of all Web security firms as it did with e-commerce stocks. As Figure 9-2 indicates, some Web security firms saw their stock prices deteriorate far more than others. Furthermore, a few Web security firms actually experienced an increase in value in 2000.

Two of the firms that were punished most severely in 2000 were V-One and VeriSign. V-One suffered because of its consistently unprofitable operations and its shrinking revenue base. Even prior to 2000, investors were willing to forgive rapid revenue growth coupled with huge operating losses but had no tolerance for declining revenues and operating losses. VeriSign, by contrast, was growing revenues rapidly. Its announcement of a $20 billion acquisition of Network Solutions came at the peak of the NASDAQ. Given the huge changes for goodwill that the acquisition would create, investors realized that the revenue gains from the merged firm were not likely to be offset by the massive writeoffs of goodwill—the difference between the $20 billion purchase price and the book value of Network Solutions' equity.

By contrast, a couple of profitable Web security firms actually saw an increase in their share prices during 2000. Check Point Software, driven by its rapid revenue growth and ample net profit margins, was up 169% in 2000. ISS Group's positive trend in operating margins, driven in part by its move into managed Web security services, contributed to investors' relatively high level of enthusiasm.

INTERNET INVESTMENT DASHBOARD ANALYSIS OF WEB SECURITY

While the foregoing discussion of the relative stock price performance of Web security segments is suggestive, it does not offer a fully satisfactory explanation. To obtain such an explanation, we apply the Internet Investment Dashboard (IID) analysis to the Web security segment.

As we will see, the IID analysis suggests that Web security could generate attractive investor returns in the future. The Web security industry clearly has a moderate amount of economic bargaining power. Hacker attacks scare decision-makers into spending money on Web security; how-

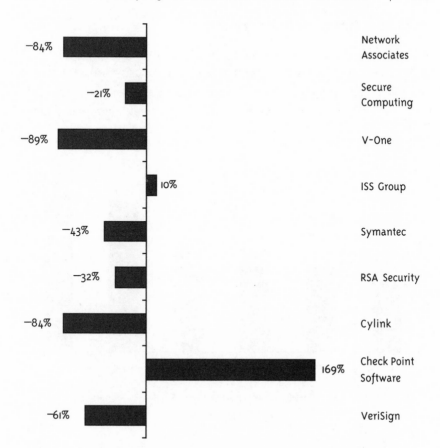

Figure 9-2. Selected Web Security Firms' Percent Change in Stock Price,
12/31/99—12/29/00
Source: MSN MoneyCentral

ever, the large number of potential suppliers puts a cap on how much vendors can charge for the services. The providers that offer customers a closed-loop solution generally have been the profitable participants. Several industry participants have demonstrated their ability to adapt effectively to change and have created a compelling brand family, and these factors help explain their superior performance in the stock market. Furthermore, there are several firms that manage their financial affairs quite effectively, and some of these companies' shares are relatively inexpensive. As a result, there could be some interesting investment opportunities within the Web security segment.

Economic Bargaining Power

Web security is a moderately large industry, growing at a moderate rate. As noted earlier, Web security consists of Web security software and services. The software segment, as illustrated in Figure 9-3, was expected to reach $4.7 billion in 2000 and is forecast to grow at a compound annual

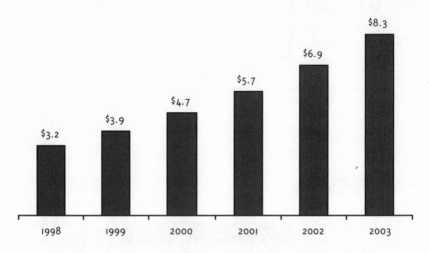

Figure 9-3. Worldwide Internet Security Software, 1999–2003,
in Billions of Dollars
Source: International Data Corp.

rate of 21%. The service segment, as illustrated in Figure 9-4, is anticipated to be about 25% of the size of the software business, growing at a 45% annual rate. As alluded to earlier in the chapter, the service business is potentially a good segue into the software. The service actually highlights the need for changes in the organization, including new processes and software to help provide network security.

Web security generates net losses for the average industry participant, but the average masks wide variations in performance among the firms analyzed. As Table 9-1 illustrates, the average net margin for the nine Web security companies sampled in this chapter is −17%. This means that for every dollar these companies generated in revenue, they lost 17 cents. In fact, this average performance includes big money losers like V-One and Secure Computing, which lost $1.86 and 73 cents, respectively, for each dollar of revenue and big profit generators like RSA Security and Check Point Software, which earned 54 cents and 48 cents, respectively, on each dollar of revenue.

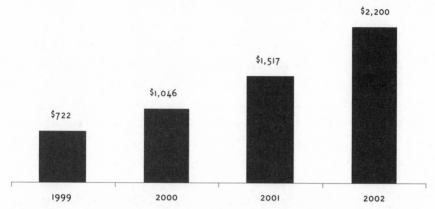

Figure 9-4. Managed Security Services Market, 1999–2003, in Millions of Dollars
Source: International Data Corporation

TABLE 9-1. MARGINS FOR NINE WEB SECURITY FIRMS, DECEMBER 2000

Company	Stock % Change	Gross Margin	Operating Margin	Net Margin	2-Year Trend
VeriSign	−61%	69%	5%	5%	+
Check Point Software	169	92	56	48	0
Cylink	−84	71	−43	−43	0
RSA Security	−32	84	88	54	++
Symantec	−43	89	39	26	++
ISS Group	10	76	12	9	++
V-One	−89	82	−186	−186	−
Secure Computing	−21	76	−73	−73	—
Network Associates	−84	87	12	6	0
MEAN	**−26%**	**81%**	**−10%**	**−17%**	**0**

Source: MSN MoneyCentral

Note: Two-year trends represent the extent to which net margins have changed in the most recent two years. ++ signifies a large improvement, + indicates a small improvement, 0 indicates no change, − indicates small deterioration, — indicates large deterioration.

Analyzing the reasons for these differences in profit can provide investors with useful insights. For V-One and Secure Computing, serious management problems have contributed to substantial operating losses and deteriorating financial conditions. By contrast, RSA Security has achieved market leadership in authentication technology, a product that enjoys significant economic leverage because buyers are sophisticated organizations that are willing to pay a price premium to protect access to their confidential information. RSA Security's particularly high profit margins are partially a result of its decision to sell some of its ownership in VeriSign, generating a substantial capital gain.

Check Point Software's market leadership in firewall technology has given the company an opportunity to extend its leadership into policy-based management—a solution that integrates authentication, encryption, firewalls, and virus protection into a closed-loop solution. By meeting customers' need for a closed-loop network security solution, Check Point Software is able to charge a price high enough to earn 92% gross margins and 48% net margins—a level that makes Check Point Software more profitable than Microsoft (with a 42% net margin).

The extent to which firms' enjoy market leadership helps explain their profitability in other ways. If a firm is perceived as a leader, it does not need to spend as much on sales and marketing as a firm that is behind. As we will explore later, sales and marketing expenses take on special meaning in the Web security industry because many firms, such as Network Associates, use distributors to sell their products. These distributors cost Web security software firms both money and market share. Distributors cost money because they demand advertising funds and a cut of the revenues. They cost market share because many distributors have significant financial and management problems, making it very unlikely that they will be able or willing to invest the resources needed to improve Web security vendors' market positions. By contrast, market-leading Web security firms that sell directly to corporate buyers tend to spend less on sales and marketing, thereby generating positive operating margins.

While it is not crucial to understand here the details of each firm's business model, it is important for investors to understand the factors that influence a Web security company's ability to exercise economic bargaining power. Three factors help increase the economic bargaining power of Web security firms: market share leadership, closed-loop solution, and interoperability (e.g., the ability of a firms' products to work effectively with other firms' products). Firms that are not market share leaders, do not offer closed-loop solutions, and whose products do not interoperate with others' have less economic bargaining power and find themselves on a downward operating trajectory from which it is difficult to recover.

Closed-Loop Solutions

As noted above, Web security firms that offer customer closed-loop solutions are generally winners. Closed-loop solutions mean that customers no longer need to purchase point products and tie them together. (Point products include authentication, encryption, firewalls, and antivirus software.) Rather than buying the best in each category of product and then integrating them, customers prefer to purchase one product from a single vendor that can deliver on the promise of Web security by tying all these capabilities together in a seamless fashion. Such a closed-loop solution is a prescription for rapid growth. Conversely, firms that have failed at such integration have grown more slowly.

One way to assess the relative effectiveness of the strategies of firms in the Web security market is to analyze the firms' relative size, revenue growth, and market capitalization. This analysis can help us gain insights into which firms are generating the greatest product market momentum and the extent to which that momentum has translated into greater stock price appreciation.

The correlation between momentum in the product markets and the capital markets is partially born out by the evidence. As Table 9-2 indicates, two firms actually increased in value in 2000. Check Point Software, which dominates the firewall market, saw its stock grow 169% in 2000. Internet Security Group, a rapidly growing and profitable security firm, increased 10% in the stock market in 2000.

Losing stock market altitude was the most common result in 2000. The stock market losers fell into two categories: second-tier players who lost momentum and leaders that overreached in their acquisitions. V-One and Cylink fall into the former category; VeriSign and Network Associates fall into the latter one.

TABLE 9-2. REVENUE MOMENTUM OF NINE WEB SECURITY FIRMS, JANUARY 12, 2001, IN MILLIONS OF DOLLARS

Company	Stock % Change	12-Months Revenue	Revenue Growth	Market Capitalization
VeriSign	−61%	$305	385%	$16,760
Check Point Software	169	353	88	19,300
Cylink	−84	70	26	94
RSA Security	−32	264	30	2,163
Symantec	−43	772	7	2,602
ISS Group	10	172	68	2,501
V-One	−89	5	−14	24
Secure Computing	−21	34	33	265
Network Associates	−84	905	48	1,084

Source: MSN MoneyCentral

Some clear patterns emerge from this analysis that could be useful for investors. Firms that offer true closed-loop solutions—e.g., those that are affirmed by positive customer feedback and a large number of partnerships—tend to grow faster. This growth is generally rewarded by faster stock price appreciation. By contrast, firms that do not deliver on the promise of a closed-loop solution tend to grow less rapidly and are punished in the stock market by lower-than-average stock price appreciation. Imbalances between relative revenue growth rates and stock price appreciation tend to be repaired—particularly during periods where investors' fear of loss overwhelms their fear of missing gains.

In order to gain meaningful insights that can explain how the differences in strategies of the Web security firms translate into different levels of return in the stock market, it is important to analyze the different strategies of the firms profiled here. Table 9-3 summarizes the key strategic differences among these firms.

TABLE 9-3. STRATEGIC DIFFERENCES AMONG NINE WEB SECURITY FIRMS

Company	Competitive Strengths/Weaknesses
VeriSign	*Strengths:* Market leadership includes 100,000 Web site digital certificates issued to organizations and 3.5 million of its digital certificates for individuals; 300 organizations subscribed to its OnSite service since November 1997; strategic relationships with AT&T, Cisco, Microsoft, Netscape, Network Associates, RSA, Security Dynamics, and Visa. *Weaknesses:* Substantial net losses likely to continue.
Check Point Software	*Strengths:* Check Point's Open Platform for Security (OPSEC) provides interoperability with 250 partners; 30% of firewall market (#1 market share). *Weaknesses:* Most significant weakness is relatively high stock valuation.
Cylink	*Strengths:* Customers include large banks, as AT&T, MCI WorldCom, and Motorola. *Weaknesses:* Pittway Corporation, a maker of alarm systems, owns 30% of the company; 45% of sales from foreign customers; poor financial performance.
RSA Security	*Strengths:* A top maker of tokens which authorize entry by PINs and random-access codes displayed on cards or tokens; customers include corporations as well as users in finance, research, and government. *Weaknesses:* Tokens are a relatively outmoded technology being replaced by digital certificates.
Symantec	*Strengths:* #2 maker of security software (Norton AntiVirus), desktop efficiency (Norton CleanSweep), and PC utility (Norton Ghost) products, which collectively account for 60% of sales; new management. *Weaknesses:* Distributor Ingram Micro accounts for nearly half of sales.
ISS Group	*Strengths:* Closed-loop solution includes network security monitoring, detection, and response software and services. *Weaknesses:* Not a clear market leader, but growing rapidly.
V-One	*Strengths:* Its main product is SmartGate, which controls user access to company networks using intranets or the Internet. SmartGate uses software or smart cards to verify user identities in remote locations.

Company	Competitive Strengths/Weaknesses
	Weaknesses: U.S. government contracts make up 60% of sales; founder James Chen owns about 22% of V-One; poor financial performance.
Secure Computing	*Strengths:* Products include Sidewinder firewall and SmartFilter, a tool developed to restrict employee access to inappropriate Web sites.
	Weaknesses: Terrible financial performance, executive turnover, Cayman Islands' ownership.
Network Associates	*Strengths:* #1 software security vendor makes antivirus (40% of sales), network management, and utility software.
	Weaknesses: Sells through direct sales and distributors like Ingram Micro and Tech Data; poor acquisition track record; poor financial performance.

Source: Company Reports, Peter S. Cohan & Associates Analysis

The foregoing analysis of Web security business strategies suggests that there is some correlation between a Web security firm's stock price performance and the robustness of its competitive strategy. For example, VeriSign and Check Point Software both enjoy market leadership and strong partnerships with leading companies in their industries. Interestingly, Symantec and Network Associates, which also enjoy very strong market shares, have suffered in their relative stock market performance, partially because of their heavy dependence on distributors such as Ingram Micro and Tech Data. As we noted earlier, dependence on distributors is a significant problem. Not only are distributors likely to have significant financial difficulties, but they also have trouble focusing on maximizing the market position of any one vendor's products.

Furthermore, the ability to distribute software directly to customers over the Internet creates a significant channel conflict problem for firms like Symantec and Network Associates. Simply put, while it might be cheaper and more effective in the long run to distribute software to customers directly over the Internet, these vendors are afraid of alienating their distributors in the medium run. As a result of this fear, these firms have trouble making the transition and remain stuck in the middle. They remain dependent on an increasingly ineffective means of distributing their product, a dependence that causes them to grow revenues relatively slowly.

As we will explore later in this chapter, the analysis of competitive strengths and weaknesses highlights an important factor that differentiates stock market winners from losers—management. Symantec suffers from many of the same strategic problems as Network Associates; however, Network Associates' previous management was significantly overpaid and mismanaged a slew of ill-conceived and poorly executed acquisitions. After years of unwarranted tolerance, Network Associates' board replaced its CEO and CFO in January 2001. Despite its $653 million in revenues, Network Associates remains an undervalued company because of its inept management. Investors could do well in Network Associates stock if the current management team were replaced with a top-notch one. It remains to be seen whether its new management can deliver.

While more quantitative measures of the effectiveness of competitive strategy, such as relative size and rapid revenue growth, are important, the quantitative factors seem to provide greater insight into which companies investors should avoid. For example, firms like Secure Computing, whose weak financial position and tiny market share led to a particularly poor showing in the stock market, are a good example of what investors should avoid. Network Associates offers a good example of a firm with problems. Even with its relatively large market shares, Network Associates is not necessarily the best company in which to invest because its weak strategies are further hampered by more qualitative factors, including botched acquisitions, conflicted distribution, and untested new management.

Investors should seek out firms that excel in the quantitative and qualitative factors driving their competitive strategies. Firms like VeriSign, Check Point Software, and ISS Group have all grown into substantial firms with rapid revenue growth. The latter two firms also enjoy profitable operations, and all of them are relatively well managed, with strong customers and partnerships. These firms are likely to continue to prosper and thus warrant further analysis.

Management Integrity and Adaptability

Management integrity and adaptability are particularly important in Web security. Customers and employees are highly dependent on Web security managers' ability to meet their commitments and to anticipate and adapt to changing technology, customer demands, and competitor strategies. If a Web security firm loses its ability to meet its commitment to employees or customers, then employees are likely to take a job at a firm that can meet its commitments and customers are likely to take their business elsewhere.

An analysis of legal proceedings and accounting policies for several Web security firms suggests that not all Web consultants have equal levels of management integrity. All the Web security firms analyzed in Table 9-4 had significant legal proceedings with the exception of VeriSign. These proceedings suggest that in many cases Web security firms' managers presided over periods of deteriorating financial performance and then did not disclose the problems in a timely way. These proceedings suggest concerns about these firms' management teams' ability both to achieve ambitious performance objectives and to deal directly with others in reporting their failure to do so.

TABLE 9-4. REVENUE ACCOUNTING POLICY AND LAWSUITS FOR WEB SECURITY FIRMS

Company	Revenue Accounting Policy	Legal Proceedings
VeriSign	VeriSign defers revenues from the sale or renewal of digital certificates and recognizes these revenues ratably over the life of the digital certificate, generally 12 months.	None
Cylink	Revenue is recognized when evidence of a sale exists, such as receipt of a contract or purchase order, product has been shipped, sales price is fixed, collection is probable, and evidence exists to allocate the fee to any undelivered elements.	Lawsuits alleging that previously issued financial statements were materially false and misleading and that the defendants knew or should have known that these financial statements caused Cylink's common stock price to rise artificially.
RSA Security	Recognizes revenue from products when shipped, no further material obligations exist, and collection is considered probable. Revenue from shipments to distributors is recognized on receipt of evidence of sale to end-users.	Lawsuits alleging that RSA misled the public concerning demand for its products, the strengths of its technologies, and trends in its business. Kenneth P. Weiss, the founder of RSA, demanded arbitration, alleging that RSA

Company	Revenue Accounting Policy	Legal Proceedings
		breached its obligations to Weiss by refusing to release assignments of patents.
Symantec	Recognizes revenue upon evidence of an arrangement, delivery of software to the customer, determination that there are no post-delivery obligations, and collection of a fixed license fee is considered probable. Defers revenue relating to all distribution and reseller channel inventory. Offers the right of return of its products.	Various lawsuits. For example, a complaint alleging that Symantec inflated its stock price and then sold stock based on inside information that sales were not going to meet analysts' expectations. Symantec sued Network Associates for infringement and unfair competition. Symantec was sued for copyright infringement by a California software company.
ISS Group	ISS recognizes perpetual license revenues from ISS-developed products upon delivery of software or, if the customer has evaluation software, delivery of the software key and issuance of license, assuming that no vendor obligations or customer acceptance rights exist.	On July 13, 1999, ISS and Network Associates announced that the patent infringement suit filed by Network Associates in July 1998 against ISS was resolved to the parties' mutual satisfaction. The resolution of this previously pending litigation had no material adverse effect on ISS business.
Secure Computing	Recognizes product revenues at the time of shipment in instances where there is evidence of a contract, the fee charged is fixed, and collection is probable.	Class-action complaints alleging false and misleading statements about its business condition and prospects
Network Associates	Records sales to distributors as revenue and at the same time establishes a reserve for returns.	Disputes and lawsuits related to securities law, intellectual property, license, contract law, distribution, and employee matters.

Source: Company 10Ks

The accounting practices of firms that deal extensively with distributors suggest room for revenue fluctuation. More specifically, Symantec and Network Associates ship products to distributors and make allowances for returns of those products. These allowances are quite typical, however, but they do raise concerns for investors who may have reason to wonder about the extent to which firms might be inclined to "stuff" their channels with products in order to meet ambitious revenue targets. Some firms are subsequently forced to adjust their revenues downward when the excess inventory is returned to inventory. While there is no indisputable evidence that such channel stuffing has occurred with these firms, the possibility of such activity could make investors question the reliability of revenue numbers from these firms—and hence to view as precarious the market capitalization keyed off of these revenues. These questions can become particularly acute in cases where shareholders have sued the firms, alleging misrepresentation of the firms' financial conditions.

While these lawsuits and accounting practices do not necessarily mean that the Web security firms have violated any laws, they do suggest that there could be some bending of limits. This bending of limits may not constitute violation of the law; however, they could constitute a violation of investor and customer trust in some cases. The impact of this environment of diminished trust could contribute to an unwillingness on the part of investors to make a bet on an uncertain commercial future.

In light of the rapid rate at which the competitive environment changes, the ability of managers to adapt to this change effectively is a crucial factor in determining the long-term value of a Web security company. To analyze this factor, we can examine the variance of the percentage change in quarterly earnings. Simply put, if a Web security firm is able to increase profits fairly consistently in each quarter, its management is good at adapting to change. Conversely, if a firm's quarterly performance jumps around significantly, its management team is probably less effective at adapting to change. The important point here is that the ability to adapt effectively to change in a rapidly changing environment can be measured by management's ability to produce fairly consistent financial results.

In the case of Web security firms, it is clear that this measure is a somewhat useful one. As Figure 9-5 indicates, consistent earnings growth is particularly important to investors in profitable Web security firms. Among the top-performing stocks was Check Point Software, which sustained profitability while achieving such consistent profit growth that the change in its quarterly EPS had a barely perceptible 2% variance. RSA Security, another strongly performing stock that generated consistent profits, had a relatively small 12% variance.

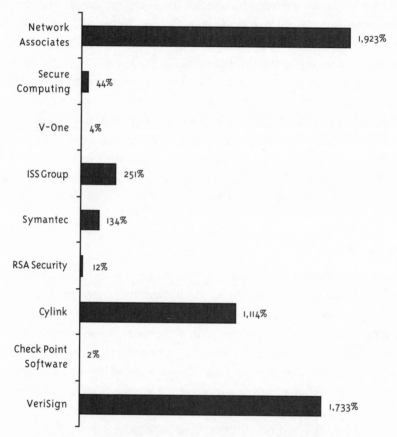

Figure 9-5. Variance in Quarterly Earnings Percent Change for Selected Web Security Firms, 3/99–3/00
Source: MSN Money Central, Peter S. Cohan & Associates Analysis

Figure 9-5 also indicates that the variance in quarterly EPS changes had no predictive value for unprofitable firms. For VeriSign, which was profitable in only one of the most recent five quarters, the variance of EPS changes was extremely high. This variance was almost as high for the unprofitable and shrinking Network Associates. Conversely, one of the worst-performing stocks, V-One, seemed to lose the same amount of money consistently every quarter, leading to a tiny 4% variance in quarterly EPS change. This outcome suggests an obvious conclusion: Consistently awful performance should not be viewed as an indicator of management adaptability.

Management integrity and adaptability are critical factors in differentiating winning Internet companies from their peers. The average Web security company analyzed here seems to have a less than outstanding performance on both of these qualitative factors. Perhaps Check Point Software has best demonstrated an ability to adapt effectively in terms of building or acquiring the skills that customers require. It is also clear that Check Point Software's adaptability has led to an extremely profitable business that delivers consistent earnings growth. Many other firms in the Web security segment have not adapted as effectively to change and have given investors reason to question their integrity. The important point for investors to use in their analysis is that management adaptability and integrity—while difficult to quantify—play an important role in the long-term value of a Web security firm.

Brand Family

Since it is often difficult for investors to understand the details of the Web security process, investors can analyze the firms' brand families as a gauge of the firms' relative value. The conclusion of the analysis of the different firms' brand families is that there are some important differences among the firms, particularly in terms of their customers and principal investors. These differences foreshadow different capabilities of shareholder value generation and are therefore useful indicators for investors.

The brand families of Web security companies vary significantly depending on whether they deal extensively with distributors. As Table 9-5 indicates, several firms that deal directly with customers rather than working through distributors seem to have created more compelling brand families. For example, VeriSign's partnerships with Microsoft, Intel, and Visa as well as its customer relationships with GE and the Internal Revenue Service suggest that VeriSign excels at its business. In addition, it appears that investors seem to place more value on firms with a mix of strong commercial customers and a few government agencies of great strategic significance. Secure Computing, which appears heavily dependent on government contracts, is much less highly valued than Axent, which has a mixture of commercial and government business.

TABLE 9-5. MAJOR CUSTOMERS AND PARTNERS OF SELECTED WEB SECURITY COMPANIES

Company	Major Customers/Partners
VeriSign	Strategic relationships with BT, Cisco, Microsoft, Netscape, RSA Security, and Visa. Website Digital Certificate services are used by all of the Fortune 500 companies with a Web presence and all of the top-40 electronic commerce Web sites as listed by Jupiter Communications. Managed certificate services subscribers include Agilent, Bank of America, Barclays, General Electric Information Systems, Hewlett-Packard, the Internal Revenue Service, Southwest Securities, Sumitomo Bank, Texas Instruments, Visa, and US West.
Check Point Software	Customers include Cap Gemini, McManus Group, Robert Mondavi Winery, *Seattle Times*.
Cylink	Customers include Fortune 1000 companies, financial institutions, government agencies, and telecommunication carriers.
RSA Security	Has strategic relationships with 500 vendors—including AOL/Netscape, Apple Computer, Ascend, AT&T, Check Point, Cisco Systems, Compaq, IBM, Intel, Microsoft, Nortel Networks, Novell, Oracle. Licensed 6 million RSA SecurID authenticators to 5,000 customers worldwide.
Symantec	Dealers and distributors are the primary means of distributing. Customers are individuals.
ISS Group	ISS' life-cycle security management products have 5,000 customers, including 21 of the 25 largest U.S. commercial banks, 9 of the 10 largest telecommunications companies, and 35 government agencies. Alliances with Check Point, GTE, IBM, MCI WorldCom (Embratel), iXL, BellSouth, Microsoft, Nortel, and Nokia.
Secure Computing	Customers include Defense Advanced Research Projects Agency (DARPA), the National Security Agency (NSA), and the Department of Energy (DOE). Commercial accounts include half of the Fortune 100.
Network Associates	Distributors include Ingram Micro, Merisel America, Pinacor, Inc., and Tech Data. Computer and software retailers include Comp USA, Staples, Best Buy, Office Max, and Office Depot. Corporate resellers, including Software Spectrum, ASAP Software, Softmart, Corporate Software & Technology, and Software House International.

Source: Company Financial Statements (10Ks)

Investors seem to place a lower value on the revenues generated by firms that sell through distributors and resellers. For example, although Symantec and Network Associates are by far the largest companies in terms of revenues, their stock price performance has been weak in comparison to firms that sell directly to end users. Network Associates' dependence on Ingram Micro, Office Depot, and Tech Data tends to dilute whatever brand strength may have been related to the specific products in its portfolio. Network Associates' software distributors are not as committed to brand development as the direct sales forces of Network Associates' competitors. The use of distributors is an important factor in limiting the relative strength of the brand families.

While the composition of major investors was a useful signal of value when the firms were first taken public, the current composition of Web security firms' investor populations is not helpful in explaining variations in stock price performance. As Table 9-6 indicates, Web security firms' ownership consists of a mixture of large money managers—primarily for retail investors—and a few insiders. This was not always the case. For example, when VeriSign went public, its potential value was signaled through large ownership positions from Microsoft, Intel, and Visa. VeriSign's most recent proxy material reveals that these firms have sold so much of their positions in VeriSign (potentially all their positions) that these signals of value are no longer part of the investor brand family.

TABLE 9-6. MAJOR INVESTORS IN SELECTED WEB SECURITY COMPANIES

Company	Major Investors
VeriSign	Putnam Investments (8.5%), Janus Capital (7.9%), D. James Bidzos (4.6%)
Check Point Software	Insiders: 30%; institutional: 51%
Cylink	Leo A. Guthart (29.40%), Paul R. Gauvreau (28.46%), Pittway Corporation (28.27%), Bermuda Trust Company (5.74%), GeoCapital (5.42%)
RSA Security	Massachusetts Financial Services Company (11.4%), Franklin Resources (6.5%) Kenneth P. Weiss (5.6%), FMR (5.0%), Charles Stuckey (1.9%)
Symantec	Mellon Bank (6.2%), J. & W. Seligman (5.6%), Legg Mason (5.4%), Gordon E. Eubanks (1.6%)

Company	Major Investors
ISS Group	Christopher W. Klaus (13.6%), Ark Asset Management (10.3%), Thomas E. Noonan (6.5%)
V-One	James F. Chen (18.0%), David D. Dawson (2.0%)
Secure Computing	Swenson Ventures (5.1%), Westgate International c/o HSBC Financial Services (Cayman) Limited (7.5%), John McNulty (1.2%)
Network Associates	William L. Larson (1.5%), HSBC Holdings Plc ($122 million), Brown Capital Management ($103 million), American Express Financial ($91 million)

Source: Company Proxy Statements

Investors place a limited emphasis on brand family when evaluating Web security firms. On the other hand, investors appear to place a higher value on Web security firms that have assembled a strong group of technology partners and customers. Investors also seem to place less value on Web security firms that sell primarily through distributors. Investors do not seem to analyze current investors in Web security firms as a means of distinguishing between valuable and less valuable firms. While this analysis helped propel VeriSign's initial stock market success, it is no longer a factor. In the future, the most important element of the brand family of Web security firms is likely to be the quality of customers and partners.

Financial Effectiveness

Prior to the second quarter of 2000, investors used financial effectiveness as a means of discriminating among Web security firms, although in a limited way. As we shall see, investors recognized and rewarded the vastly superior financial effectiveness of Check Point Software both before April 2000's Internet stock crash and, in a different way, after that crash.

More specifically, investors placed more emphasis on revenue growth and market leadership prior to April 2000. As Figure 9-6 suggests, the differences in sales productivity among Web security firms was not an important factor that investors used in driving relative stock price performance. For example, the sales productivity of VeriSign—the top stock market performer—was only a bit higher than the sales productivity of the worst stock market performer in the group—Network Associates.

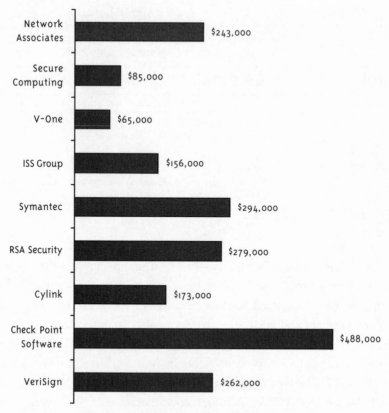

Figure 9-6. Sales per Employee of Selected Web Security Firms, 1999
Source: MSN MoneyCentral

Nevertheless, there are some distinctions in stock market performance that the sales productivity analysis helps explain. For example, two of the worst stock market performers—V-One and Secure Computing—generated much lower sales productivity than their peers.

These firms' stock prices also held up much less well during the April 2000 Internet stock crash. Conversely, Check Point Software, which generated by far the highest sales productivity in the group, declined far less than VeriSign did during April 2000. Simply put, after April 2000 investors seemed to believe that financially effective Web security firms were better stores of value than those that were not.

The other measure of financial effectiveness, the SG&A-to-sales ratio, tells the same story. As Figure 9-7 indicates, V-One and Secure Comput-

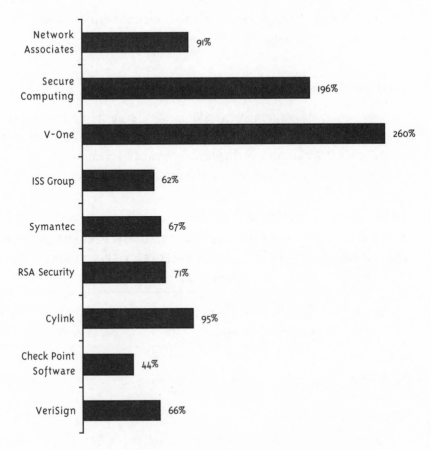

Figure 9-7. Selling, General, and Administrative Expense as a
Percent of Sales for Selected Web Security Firms, 1999
Source: MSN MoneyCentral

ing—whose SG&A exceeds their sales—have seen their stock prices plummet since April 2000. Check Point Software, which has a much lower ratio of SG&A to sales, has held up much better since April 2000. For the other Web security firms, comparing SG&A to sales was not useful for explaining stock performance.

The conclusion from this analysis is that investors in Web security firms have changed the level of emphasis they placed on financial effectiveness. Prior to April 2000, financial effectiveness did not seem to explain variations in stock price performance. After April 2000, financial effectiveness was useful in helping investors decide which marginal performers to dump and in picking those Web security firms that were more likely to survive—and thus be a good store of value for investors.

Relative Market Valuation

The stock market's approach to valuing Web security companies suggests a number of potentially over- and undervalued firms as of this writing. As Table 9-7 indicates, on the basis of price/sales ratios, for example, the firms in our sample fall into two categories: double-digit and single-digit.

TABLE 9-7. STOCK PRICE AND REVENUE GROWTH PERCENT CHANGE, 2000, AND VALUATION RATIOS FOR SELECTED WEB SECURITY FIRMS, JANUARY 12, 2001

Company	Stock % Change	Revenue Growth	Price/Sales	(Price/Sales) /Rev. Growth	Price/Book
VeriSign	−61%	385%	54.76	14.22	0.85
Check Point Software	169	88	54.63	62.08	42.76
Cylink	−84	26	1.34	5.15	1.32
RSA Security	−32	30	8.20	27.33	4.09
Symantec	−43	7	3.37	48.14	3.76
ISS Group	10	68	14.60	21.47	14.08
V-One	−89	−14	5.45	−38.93	4.95
Secure Computing	−21	33	7.80	23.64	12.07
Network Associates	−84	48	1.20	2.50	1.55
MEAN	**−26%**	**75%**	**16.82**	**18.40**	**9.49**

Source: MSN MoneyCentral

The double-digit firms include VeriSign, Check Point Software, and ISS Group. Within the double-digit firms, two firms may be overvalued (Check Point Software and VeriSign). Check Point Software is an exceptionally strong performer both in terms of market share and financial performance. Compared to VeriSign, it is likely that Check Point Software is undervalued on a price/sales basis. However, relative to peers such as ISS Group, it is not clear that, at 55 times sales, Check Point Software is worth three times as much as ISS Group, which is growing at 68%.

Of the single-digit firms, V-One and Secure Computing are in such rough shape financially that they should have the lowest price-sales ratios of the group. ISS Group appears to be priced to reflect its rapid growth and relatively skimpy profit margins. Finally, RSA Security, with its strong profitability and rapid growth, appears to have been too heavily punished for its reliance on token technology, which, while somewhat outmoded, appears to continue to be popular among the sophisticated organizations that use it.

Other single-digit firms—Cylink, Symantec, and Network Associates—all seem to be lagging the industry in terms of growth and profitability. In light of their relatively significant revenue base, Symantec and Network Associates are in a different category than Cylink. The former firms have a strong market position that might be more fully developed under new management. More specifically, if Symantec and Network Associates could bypass distributors and sell directly to consumers over the Internet, then their cost structures might be able to drop and their profits could spurt. Such a radical transformation in distribution strategy would make these two firms clearly undervalued assets. The relatively low price-sales ratios of these firms suggests that the market recognizes the difficulty of turning this concept into a profitable reality.

Cylink faces a different challenge. Lacking the scale or market leadership of other firms in the industry, Cylink appears to be a potentially attractive acquisition target. It is possible that Cylink will be a more attractive target should its market price decline more. It may also be that the strategic benefits to potential acquirers are not sufficiently high to offset the purchase prices for Cylink. As a result, investors may not wish to speculate in Cylink shares.

The relative market valuation analysis of Web security companies suggests significant differences among firms in the industry. There could be investment opportunities to buy potentially undervalued firms such as RSA Security. It may be worth investors' time to scrutinize the potentially high valuations of VeriSign, V-One, and Secure Computing to assess whether profit could be earned by selling these shares. Should firms such as Symantec and Network Associates undergo a significant change in distribution strategy that results in lower costs and rapid profit growth, investors may be able to profit from purchasing the shares. Despite these suggestions, investors should view relative valuation analysis as one component that should be considered along with the other five components reviewed earlier in the chapter.

The Internet Investment Dashboard analysis suggests that Web security could be an attractive place for long-term investors who are selective. In Figure 9-8, all the indicators are dotted. Such a configuration strongly argues for a careful company-by-company analysis.

These indicators suggest that investors should approach the Web security sector with caution. The industry is moderate in size and growth. Profit margins vary widely by participant. Industry leaders such as Check Point Software and RSA Security enjoy good margins. Industry laggards are shrinking unprofitably. Web security is an industry where providing a closed-loop solution is particularly important. A few firms offer such solutions, while many sell point products or collections of point products that are not integrated. Management integrity and adaptability are also impor-

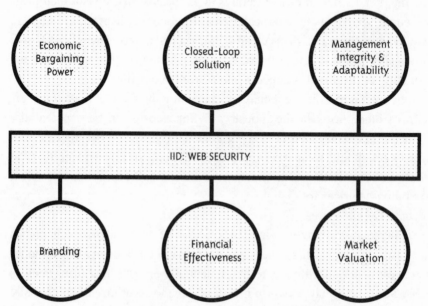

Figure 9-8. Internet Investment Dashboard Assessment of Web Security Segment
Key: Dotted = mixed performance and prospects.

tant in evaluating Web security firms. Very few firms pass this third test. Brand families are important for Web security firms. Web security firms with strong partners and well-regarded customers do better than firms that depend heavily on resellers and distributors. Financial effectiveness is increasingly important to investors seeking capital gains and a store of value. Finally, relative market valuations indicate that some Web security firms appear overvalued and others potentially undervalued.

Web security offers careful investors significant opportunities. For example, using many of the indicators described in this chapter, in August of 1998 Check Point Software was an interesting stock to buy at a price of $5.50 per share. As of December 2000, Check Point Software was trading at $134. Whether such Web security 23-baggers will exist in the future depends on how the six IID factors evolve.

PRINCIPLES FOR INVESTING IN WEB SECURITY STOCKS

The analysis we have conducted here is subject to change. Some of the strong-performing companies will falter, some of the weak ones could generate positive surprises, and new competitors could emerge to tilt the playing field in a totally different direction. As a result, investors in Web security stocks need some principles to guide them through the analytical process as investment conditions evolve.

Web Security Investment Principle 1. Identify clusters of Web security products and services with economic leverage. As we noted earlier, the average Web security firm is unprofitable. But this average masks wide variations among the firms based on the specific clusters of Web security products and services they provide customers. Investors should look for firms or groups of firms that offer Web security products and services that are so valuable to organizations that these organizations are willing to pay a significant price premium to obtain these clusters of products and services. One such bundle is secure-payment networks—a combination of consulting services to identify network breaches in payment and the provision of software to patch those breaches and provide secure payments. Such product/service bundles are likely to be of strategic value to firms selling goods over the Internet. As a consequence, firms providing such payment networks are anticipated to be able to charge a price that enables them to earn substantial profits as the number of network users grows to a significant scale. Investors who can identify such profitable product/service bundles are on their way to identifying profitable investment opportunities in the Web security segment.

Web Security Investment Principle 2. Monitor firms within the profitable Web security segments to pinpoint those that offer closed-loop solutions whose value is verified by customers and independent analysts. As we have noted repeatedly throughout this chapter, Web security customers place a significant value on receiving a closed-loop solution. These customers do not generally wish to purchase individual Web security point products and integrate them. The challenge for the investor is to pinpoint the specific Web security vendors that actually deliver on the promise of the closed-loop solution. Since several industry journals conduct independent testing of Web security products and services, studious investors can get the help they need to distinguish among the vendors.

Web Security Investment Principle 3. Avoid investing in firms where there are questions about management integrity and adaptability. Our analysis indicated questions about management integrity and adaptability in a few firms. This analysis is most useful for investors as a sell signal. If a firm with a track record of avoiding shareholder lawsuits and generating consistent profit growth suddenly deviates from this path, there is a good chance that investors should consider selling the stock. In many cases, such deviations from the path of integrity and adaptability could be a signal that management is no longer up to the job. When these deviations occur, it is likely that management will try to gloss over the problems. Investors may not want to stay along for the ride.

Web Security Investment Principle 4. Use compelling brand families as a signal of value in conjunction with other indicators. As we noted several times in this chapter, the quality of a Web security vendor's partners and customers sends a strong signal about the firm's relative value to potential investors and future customers. It is highly likely that a Web security firm that starts off in business with highly regarded customers and partners will be able to sustain this level of quality. Firms that start off with strong partners and customers and later allow the quality of their brand families to deteriorate may be sending a signal to investors that their shares should be sold. Conversely, firms whose partners and customers gain in quality over time may be signaling an increase in their share value, which could make them good investments.

Web Security Investment Principle 5. Use financial-effectiveness measures to distinguish among long-term leaders, medium-term survivors, and liquidation candidates. While investors seem to have tolerated the absence of financial discipline among Web security firms up until April 2000, this patience is fading. Firms that lack financial discipline are unlikely to survive as independent entities unless they can obtain outside financing, and this is becoming more difficult. Firms that generate strong profitability and rapid revenue growth are likely to be rewarded with more highly valued stock, which they can then use as currency to acquire valuable assets. Such currency can contribute to a virtuous cycle of profitable growth if it is used wisely. If not, as in the case of Network Associates, it can create corporate zombies that continue to survive but fail to grow and prosper. While there may be a set of companies for whom such considerations are not important determinants of stock price performance, investors should seek out Web security firms whose financial effectiveness is improving and avoid those whose financial effectiveness is eroding.

Web Security Investment Principle 6. Monitor relative valuations among Web security sectors and within Web security sectors to identify potentially under- and overvalued companies. As we noted earlier, investors seem to make broad valuation distinctions among Web security firms partially on the basis of their distribution channels. Firms that sell directly to organizations tend to have higher price-sales ratios than those that sell through distribution channels. Within these two broad categories, valuations can be distinguished on the basis of relative revenue growth and profitability to identify firms that may be under- or overvalued. For example, Check Point Software appeared to be grossly undervalued in the fall of 1998, after it had fallen from $60 to $20 even as it posted doubling of revenues, tripling of profits, and 40% net profit margins. This mismatch between dropping stock price and exploding operating performance proved to be a rare and valuable investment opportunity.

CONCLUSION

Web security presents opportunities to the discerning investor. Investors who can pick out clusters of Web security products and services that are valuable to powerful decision-makers must then find individual firms that can deliver on the promise of the closed-loop solution. If those firms are managed by great teams who adapt well to change and are skilled at signing up prestigious customers and partners, then investors should analyze their financial effectiveness and look to buy shares in these firms when their relative valuations are low. While such screening may be arduous, the additional effort is likely to yield good investment returns in the Web security sector.

Chapter 10

PROFIT-FREE NAVEL GAZING:
WEB CONTENT
STOCK PERFORMANCE DRIVERS

THE MEDIA SEEMS OBSESSED with the Internet even as it tries to figure out what to do about the Internet in the media business. So far, no publicly traded firm has succeeded in making Web content a profitable business, even with the growth that has accompanied other Internet business segments. As a consequence, there are many sources of information about Web business whose content is quite interesting for readers but whose stock prices are likely to continue to remain in the doldrums. Unless some firms can figure out how to evolve into profitable growers, investors should steer clear of must companies in the Web content segment.

Forrester Research, a publisher of research reports on Internet business, has figured out how to make a profit in the Web content business, and the stock market has rewarded it accordingly. Forrester Research's stock has appreciated 244%, from $16 in September 1999 to $55 in September 2000. Ironically, while the reasons for Forrester's stock market success—a strong financial position, profitability, strong management team, and good products—do not make interesting news copy, they have helped to propel the stock upward.

By contrast, TheStreet.com, whose founder, James Cramer, has a penchant for making news himself, has punished stockholders. TheStreet. com's stock lost 74% of its value in one year, from September 1999 to September 2000, tumbling from $27 to $7. TheStreet.com seemed to have many advantages, including the backing of The New York Times Company and Goldman Sachs. Unfortunately for shareholders, TheStreet.com's prestigious pedigree has not helped turn a fundamentally money-losing business model into an attractive stock. Nor has management turnover—such as the April 2000 departure of its CFO—helped with the stock market performance. The August 2000 infusion of $7.5 million from Paul Allen's Vulcan Ventures was not sufficient to stanch the downward movement of TheStreet.com. By November 2000 TheStreet.com announced a 20% workforce reduction, the closing of its British operation, and the termination of its joint newsroom with the *New York Times*. The future remains cloudy for Web content firms like TheStreet.com.

As these examples suggest, many organizations are trying to make a living by producing content about the Internet. These Web content firms are providing much valuable information and analysis through a variety of media. Some Web content firms produce research reports and sponsor conferences that attract organizations that purchase and use Web products and services. Other Web content firms generate advertising revenue by posting banner ads targeted at individuals who visit the Web sites to read free information and analysis. The challenge for investors is that people find value in Web content but are generally uncomfortable paying enough money for that content to offset the costs of producing it. Hence, most Web content firms are unprofitable and have made lousy investments.

Web content firms face three problems. The first problem is that people have been conditioned to receive information free, so it is difficult to charge subscriptions. The second problem is that writers of the content have been conditioned to receive low pay and are often uncomfortable with the concept of profit. The third problem is that banner advertising has a very low click-through rate (about 0.6% of banner ads are clicked-through). As a result, the payoff on banner advertising is very low, and therefore the amount that an advertiser is likely to pay for this advertising is low as well.

WEB CONTENT STOCK PERFORMANCE

The stock price performance of the Web content segment has dragged down the Internet stock average. We noted in Chapter 3 that the stock market performance of a cross section of Web content firms was 22% in 1999. This average performance is much lower than the NASDAQ's 86%

increase in 1999 and even worse than the 339% average stock price increase of the companies in our nine-segment Internet stock index in 1999. In 2000, Web content stocks lost 47% of their value, a two-year net negative return of 33%.

The lagging performance of Web Content stocks would be even worse if it were not for the strong performance of one firm, CNET, which was one of the first Web content firms. Based in San Francisco, CNET built many distribution channels for its content, including TV, radio, print, and the Web. Ultimately, CNET got out of the TV business, sold its Web portal, and focused on generating revenues from advertising and e-commerce. CNET was seen by technology experts as an excellent destination for researching the purchase of technology products such as PCs. After visiting CNET to decide which PC to purchase, users would then visit the Web site of the vendor of the selected PC. When CNET began to take a commission for generating these leads, it began to earn a profit. This transition to a profitable firm caused the stock market to take an interest—driving its stock price up 326% in 1999.

Without CNET's strong performance, Web content firms would have lost investors' money in 1999. As Figure 10-1 illustrates, the price of most Web content stocks dropped in 1999. The best performers were the leading Internet research firms, such as Forrester Research. These firms were profitable and had such a high profile in the media that investors believed that there might be sufficient value in their shares to warrant some upward movement, albeit much less than the overall Internet stock average. The worst 1999 Web content performers were Web content sites TheStreet.com and CBS MarketWatch. While the quality of their content was high, the cash flow from their operations was decidedly negative.

The effect of the April 2000 crash in Internet stocks had a somewhat unusual effect on the Web content firms' stock prices. Unlike the e-commerce stocks, the market did not uniformly destroy the value of all Web content firms. As Figure 10-2 (page 251) indicates, some Web content firms saw their stock prices deteriorate far more than others. Furthermore, a few of these firms actually experienced an increase in value in 2000.

Two Web content firms that were punished most severely in 2000 were CNET and TheStreet.com. CNET suffered in the stock market because it decided to spend $100 million to advertise its brand more heavily. This advertising campaign turned CNET from a profitable firm to an unprof-

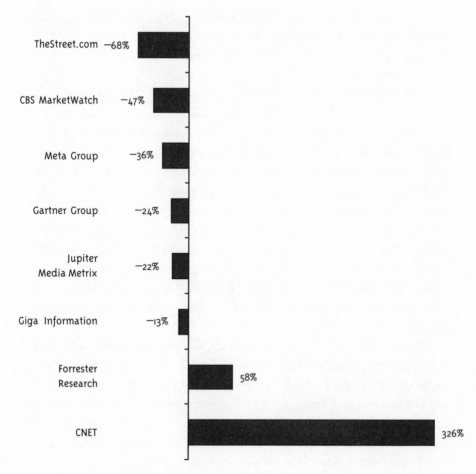

Figure 10-1. Selected Web Content Firms' Percent Change in Stock Price,
12/31/98—12/31/99

itable one. This short-term reversal was considered a temporarily unpardonable sin by investors, who knocked 72% off the value of the stock in 2000. TheStreet.com suffered from more severe problems, including a massive cash burn rate and significant management turnover. As investors knocked an additional 85% off the value of TheStreet.com's stock in 2000, the company was rumored to be seeking a buyer.

Two Web content firms, Forrester Research and Giga Information, actually saw a fairly significant increase in their stock prices during 2000. Forrester Research operated profitably; Giga Information consolidated its conferences business and discontinued the operations of its BIS Information unit, which it had acquired several years ago. These cost-reduction

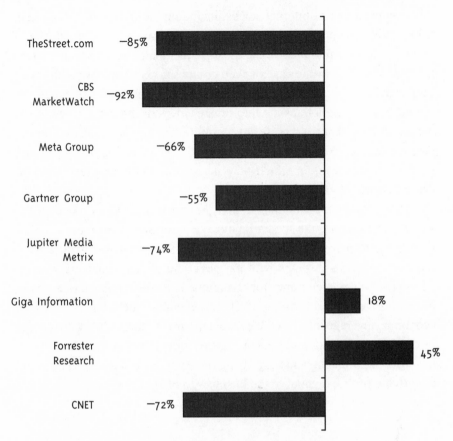

Figure 10-2. Selected Web Content Firms' Percent Change in Stock Price,
12/31/99–12/29/00
Source: MSN MoneyCentral

measures, coupled with a 40% increase in revenues, convinced investors that Giga Information had the potential to recover from what appeared to be a certain corporate death.

The stock market behavior of Web content firms underscores the curious power that the media has over stock prices. When a media company, particularly a Web media company, sells its shares to the public, the Heisenberg uncertainty principle kicks in. Paraphrased, this principle implies that the act of measuring a phenomenon changes the nature of the phenomenon itself. How does this principle relate to the stock market behavior of Web content firms?

The media, including the Web content firms, are shaping perceptions of the future value of Internet stocks, including the stocks of Web content firms. This perception-shaping process causes investors to buy stocks in the firms that seem to receive the most frequent and most favorable attention in the media. In the case of Web content firms, frequent mentions of Forrester Research and Jupiter Media Metrix, for example, as experts in the Internet business translate into greater investor interest in these companies' stocks. Since a significant number of individual investors do not take the time to examine the financial performance and industry position of many of the firms in which they invest, the media coverage tends to drive their investment decisions.

The irony of the situation is that the media also loves blood in the water. For example, the media enjoys covering problems with its peers, giving ample play to TheStreet.com's cash flow problems, management turnover, business disputes with partners, and ethical problems with its CEO's media comments and stockholdings. The effect of this coverage has been to drive down the value of TheStreet.com's stock. Given the weak economic leverage of many Web content firms, their ability to charge prices for advertising and content partnerships is not sufficient to offset the costs of producing and advertising the content. As a result, there is not underlying profit potential in the business to offset the impact of unfavorable media coverage on the stock.

INTERNET INVESTMENT DASHBOARD ANALYSIS OF WEB CONTENT

While the foregoing discussion of the relative stock price performance of Web content segments is suggestive, it does not offer a fully satisfactory explanation. To obtain such an explanation, we apply the Internet Investment Dashboard (IID) analysis to the Web content segment.

As we shall see, the IID analysis suggests that Web content is unlikely to generate attractive returns for investors in the average companies in the industry. The Web content industry clearly has very little economic bargaining power. Web content users are generally unwilling to pay for information. Advertisers are not spending enough money on Web content sites to offset the cost of producing the content. Web content firms do a great job of providing content for readers, but they are less effective at providing a compelling return on investment for advertisers. While many of the Web content firms have created compelling brand families, they have not adapted particularly effectively to change. Most of the Web content firms have done a poor job of financial management, and the stock market seems to have applied lower comparative valuation levels to Web content firms. As a result, investors must look hard to find good investments in the Web content segment.

Economic Bargaining Power

One of the first challenges faced by potential investors in Web content firms is the difficulty of finding a market whose economic bargaining power can be evaluated. As we noted earlier, Web content, firms generally do not pay for the content. In addition, while many people visit Web content sites, they do not pay for the content, either. On the other hand, the Web content firms that sell research reports and sponsor conferences do generate actual revenues from their content. Ironically, these research firms have not produced research reports that quantify the size, growth rate, and trends in their own industry. As a result, it is difficult to quantify the size of the market from which they derive their revenues. While simply adding up the revenues of the publicly traded research firms might yield an indication of the size of the market, it would probably be an inaccurate gauge of the market. First, it would exclude the revenues of many research firms that are not publicly traded. Second, it would include revenues generated by selling research on industries other than the Internet.

One area where some of the Web content firms are seeking revenues is Internet advertising. Firms such as CNET, TheStreet.com, and CBS MarketWatch are seeking a share of these Web advertising revenues. In pursuit of these revenues, these Web content firms are competing with Web portals and e-commerce firms as well. As Figure 10-3 indicates, the market for online advertising is moderate and is expected to grow at a 51% compound annual rate. With the competition for the relatively limited online advertising budgets, and with the low click-through rates of such advertising, it is not likely that Web content firms are going to be able to negotiate high advertising rates. Unless Web content firms can find a way to attract a market-leading share of site visitors—an accomplishment that none has heretofore managed to achieve—then Web content is likely to continue to suffer from an absence of economic bargaining power.

Web content generates net losses for the average participant; however,

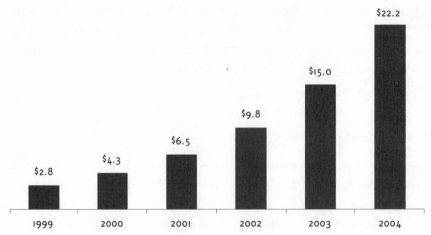

Figure 10-3. U.S. Online Advertising Spending, 1999–2004,
in Billions of Dollars
Source: DoubleClick 10K

the average masks wide variations in performance among the firms analyzed. As Table 10-1 illustrates, the average net margin for the nine Web content companies sampled in this chapter is –62%. This means that for every dollar that these companies generated in revenue, they lost 62 cents. This average performance includes big money-losers like CBS MarketWatch and TheStreet.com, which lost $1.80 and $1.97, respectively, for each dollar of revenue, and profit generators like Forrester Research and Meta Group, which earned 14 cents and 4 cents, respectively, on each dollar of revenue.

TABLE 10-1. MARGINS FOR EIGHT WEB CONTENT FIRMS, DECEMBER 2000

Company	Stock % Change	Gross Margin	Operating Margin	Net Margin	2-Year Trend
CNET	–72%	79%	19%	–40%	–
Forrester Research	45	74	20	14	0
Giga Information	–18	60	–16	–16	+
Jupiter Media Metrix	–74	70	–84	–84	–
Gartner Group	–55	58	7	3	–
Meta Group	–66	51	10	4	–

Company	Stock % Change	Gross Margin	Operating Margin	Net Margin	2-Year Trend
TheStreet.com	−85	57	−197	−197	−
CBS MarketWatch	−92	72	−180	−180	−
MEAN	**−47%**	**65%**	**−53%**	**−62%**	−

Source: MSN MoneyCentral

Note: Two-year trends represent the extent to which net margins have changed in the most recent two years. + indicates a small improvement, 0 indicates no change, − indicates small deterioration.

Analyzing the reasons for the differences in profit can provide investors with useful insights. The biggest money-losers are the Web content sites that do not charge users for their information. Firms like CBS MarketWatch are spending significant amounts of money to produce content that they give away to users free, hoping to make up the difference by cluttering their sites with paid advertising that most visitors ignore—just as they try to ignore most TV advertising. In addition, these Web content firms also spend money on advertising their sites. Given their lack of economic leverage, it is difficult to imagine a scenario in which these firms will be in a position to command prices that exceed their costs.

The firms that earn profits are the research firms that charge thousands of dollars for their research reports. Since they hire sales forces to convince deep-pocketed corporations to buy their reports, they are skilled at creating the perception that the insights to be gained from reading the reports will far offset their cost. Furthermore, by creating a fairly steady stream of new reports, these firms can encourage companies to sign up for annual subscriptions. Such annual subscriptions force their sales forces to both encourage new companies to sign up for their services and to persuade existing customers to renew their previous subscriptions. While the availability of free Web content has created internal pressure on Web research executives to cut prices, the results to date suggest that these firms have found ways to maintain their competitiveness without eliminating their profit margins.

Why can the research firms charge for their research while other Web content firms cannot? The simple reason is that the research firms are more likely to write favorable research reports on firms that subscribe to their services. Firms that do not subscribe are less likely to get the attention of the analysts who write the reports. With that lack of attention comes a lower probability that the firms will receive favorable coverage in the reports. The reports are important because they influence the purchasing behavior of the deep-pocketed corporations that buy the technology products. If a new technology company gets a favorable evaluation in a research report, that firm is likely to get more customers than a firm that gets unfavorable coverage or no coverage at all. This is the source of economic bargaining power that enables the well-established research firms— e.g., the ones that are most frequently cited in the mainstream business press— to charge high prices for their research.

The reason that investors may shy away from investing in Web content companies is that their ability to charge a price that exceeds their costs is not likely to rise. As a consequence, the only way that the average net margin of −62% of publicly traded Web content firms is likely to increase in the short term is if some of the most unprofitable firms in the segment are sold or simply liquidate their assets. The forces likely to reduce Web content firms' economic bargaining power are greater than the forces likely to increase that power. More specifically, even the most powerful research firms are threatened by the availability of free analysis of technology companies available on Web sites like CNET's. Simply put, if corporate customers can get the same level of analysis for free on CNET, they will not renew their contracts with the research firms.

The huge losses generated by some Web content firms raises the question of why investors funded them in the first place. Examining TheStreet.com and CBS MarketWatch, we realize that both firms are owned by deep-pocketed investors in other businesses. For example, TheStreet.com is owned by James Cramer, the owner of a hedge fund. It would not be surprising to discover that the level of media visibility associated with his activities for TheStreet.com would generate new customers for his hedge fund. Or it may be that Cramer simply enjoys running a media company regardless of whether it loses money. More likely, he thought it would be fun to run an online media company, but as the problems associated with that activity become onerous, he may become less enamored of the idea.

For CBS, the idea of trading TV advertising for an equity stake in CBS MarketWatch seemed clever at the time. It is possible that the value of CBS's investment is higher than the original cost—albeit much lower than it was before the stock price began to tumble off its high in 1999. It is difficult to imagine this stock ever returning to its all-time high.

Closed-Loop Solution

As noted above, Web content firms have varied levels of success at delivering closed-loop solutions to customers. In the case of Web-based content firms, the solutions provided to site visitors are very good, but the solutions for the paying customers—the advertisers—are not so compelling. As mentioned earlier, Web advertising has a 0.6% click-through rate. While the Web is a medium where it is possible to measure accurately the level of response from advertising spending, the results of this measurement are not compelling. Furthermore, with cash flow tight for many Internet firms, the rate of Web advertising spending from these firms on the Web content sites is likely to decline. In short, the Web-based content firms offer a closed-loop solution to visitors, not advertisers.

As for the Internet research firms, the nature of the solution is more powerful. The Internet research firms provide visibility to technology companies and analysis to corporate technology buyers. As long as the corporate technology buyers recognize that the research firms are likely to favor technology companies that subscribe to their services, then the research may be of value to the corporate technology buyers. For the technology companies, a favorable analysis by the most prominent research firms is well worth the cost of the subscription.

One way to assess the relative effectiveness of the strategies of firms in the Web content market is to analyze the firms' relative size, revenue growth, and market capitalization. This analysis can help us gain insights into which firms are generating the greatest product market momentum and the extent to which that momentum has translated into greater stock price appreciation.

The correlation between momentum in the product markets and in the capital markets is partially borne out by the evidence. As Table 10-2 indicates, faster-growing Web content firms above a certain size tend to perform better in the stock market than Web content firms whose revenues are lower. Forrester Research, with $137 million in revenues and growing at an 83% annual rate, saw its stock price increase 45% in 2000—the best performer of the Web content firms. Despite its rapid revenue growth, CNET stock lost 72% of its value in 2000, due to its big net loss and its dependence on a slowing Internet advertising market.

TABLE 10-2. REVENUE MOMENTUM OF EIGHT WEB CONTENT FIRMS, JANUARY 12, 2001, IN MILLIONS OF DOLLARS

Company	Stock % Change	12-Months Revenue	Revenue Growth	Market Capitalization
CNET	−72%	$191	107%	$1,950
Forrester Research	45	137	83	1,080
Giga Information	18	67	39	27
Jupiter Media Metrix	−74	47	202	271
Gartner Group	−55	859	17	679
Meta Group	−66	119	26	53
CBS MarketWatch	−92	48	162	63
TheStreet.com	−85	24	107	67

Source: MSN MoneyCentral

The 2000 stock market subtracted value from most of the other Web content stocks covered in this chapter. Stocks that dropped in value tended to be either much smaller or growing more slowly than these two firms. With $859 million in revenues, Gartner Group was by far the biggest firm in the Web content sample; however, its mere 17% annual growth rate drove investors to subtract 55% from its 2000 stock market value. Conversely, smaller firms like CBS MarketWatch, with $48 million in revenues and 162% annual growth, lost 92% of their value in 2000. Simply put, in 2000 investors rewarded companies whose scale was moderate, whose growth rates were very high, and who earned a profit.

In order to gain meaningful insights that can explain how the differences in strategies of the firms translate into different levels of return in the stock market, it is important to analyze the different strategies of the Web content firms profiled here. Table 10-3 summarizes the key strategic differences among selected firms profiled in this chapter. The table suggests that relative number of clients is not a good predictor of stock market performance. As we indicated, Gartner Group, which has the most clients of the research firms, had among the weakest stock market performance. This relatively weak performance was a result of its slow revenue growth rate. The slow rate of revenue growth was most likely a consequence of its slow rate of adaptation to covering the Web.

TABLE 10-3. STRATEGIC DIFFERENCES AMONG 8 WEB CONTENT FIRMS

Company	Competitive Strengths/Weaknesses
CNET	*Strengths:* Expanding from technology-oriented consumers through CNET.com, CNET Builder.com, and CNET Computers.com to general public with acquisition of mySimon.com (online comparison shopping); multiple content-delivery channels, including CNBC and CNET Radio. *Weaknesses:* Operating losses; change in management could be a challenge; winning in consumer market could be difficult.
Forrester Research	*Strengths:* 1,200 corporations and technology vendors as clients; 25 research services; Forrester Forum Series of conferences. *Weaknesses:* Competitive threat from free technology analysis on sites such as CNET.
Giga Information	*Strengths:* Charges an annual fee for its two Web-based continuous information services (Advisory Services and ePractices), which cover application development, networking, infrastructure, and e-commerce strategy and development; new management team. *Weaknesses:* Operating losses; weak stock price.
Jupiter Media Metrix	*Strengths:* 500 clients include America Online and IBM; Strategic Planning Services provide analysis and data about the Web only; Media Metrix market share data. *Weaknesses*: Potential slowdown in growth if Web company consolidation continues; integration challenges.
Gartner Group	*Strengths:* 9,600 organizational clients; owns 37% of Jupiter Communications. *Weaknesses:* Slow growth; management turnover; low stock price.
Meta Group	*Strengths:* 5,000 subscribers and a 78% renewal rate. Clients pay an average of about $16,700 a year for services. *Weaknesses:* Not a market leader; weak financial performance; poor stock market performance.
CBS MarketWatch	*Strengths:* Real-time news, IPO reports, mutual fund data, and personal finance tools; subscription-based services such as MarketWatch RT (real-time financial data) and MarketWatch LIVE (real-time quotes from U.S. exchanges).

Company	Competitive Strengths/Weaknesses
	Weaknesses: Poor financial and stock market perform-ance; CBS may not be patient investor.
TheStreet.com	*Strengths:* 100,000 subscribers; deals with Yahoo and AOL; Softbank owns stakes.
	Weaknesses: Poor financial performance and stock market performance; ongoing layoffs possible if New York Times selling its stake.

Source: Company Reports, Peter S. Cohan & Associates Analysis

This analysis also suggests that many of the Web-based content firms are doing a good job of assembling strategic partnerships to deliver their useful information to a broader audience of visitors. For example, TheStreet.com's partnership with Yahoo helps bring content to an audi-ence much larger than TheStreet.com's 100,000 subscribers. CNET's TV deal with CNBC similarly broadens the audience for CNET's content. These market-broadening partnerships make strategic sense. If they suc-ceed in increasing the number of site visitors, the potential advertising revenues can increase as well. However, these strategies also have signifi-cant limitations. For example, we have already noted that it is difficult for Web content firms to generate sufficient advertising to offset the costs of producing content. If these partnerships can bring in a large-enough group of additional visitors to the sites, then it is possible that the relatively small costs of making this content available for a broader audience may be offset. The problem is that it has yet to be demonstrated that these part-nerships actually deliver the hoped-for numbers of new visitors or sub-scribers.

The implications of this analysis for investors are clear. Most Web con-tent firms have yet to crack the code for creating a rapidly growing, prof-itable business that is sufficiently large to give investors confidence in its long-term viability. There are currently small, rapidly growing firms that generate huge losses while providing closed-loop solutions to visitors— not necessarily advertisers. There are large firms that are growing quite slowly because they have not completely embraced the Internet. There are a few firms that have reached sufficient scale, are growing quickly, and have achieved some profitability. Firms in the latter category may be worth further evaluation by investors seeking stock price appreciation.

Management Integrity and Adaptability

Management integrity and adaptability are particularly important in Web content. If site visitors perceive that the content on the sites is corrupted by conflicts of interest on the part of authors, then they may stop visiting the site. If employees and partners perceive that they are being mistreated, they will no longer contribute their work. If investors believe that they are being misled, they will sell the stock of the Web content firms.

An analysis of legal proceedings and accounting policies for Web content firms suggests that Web consultants have a high degree of management integrity. As Table 10-4 indicates, none of the Web content firms analyzed in this table had significant legal proceedings. The accounting policies for revenues suggest a high degree of consistency across firms. One issue that investors might want to pay attention to is the research firms' practices regarding the recording of deferred revenues and the conversion of billings into revenues. More specifically, many of the research firms account for renewed research contracts as deferred revenue—taking the billings into revenues over the period of the contract. If firms recognize revenues as the services are performed, then the revenue amounts are likely not to be overstated. Furthermore, if firms are able to collect payment on the revenues at the same time that they are recording revenues, then there is not likely to be a shortage of cash collections.

Any disconnect between the recording of revenue and the timing of cash collection would be important to investors. This disconnect would be particularly important if the revenues exceeded the cash collections and the slower cash collections forced the firms to borrow money to pay short-term obligations. Such a cash collection gap could be a sign of weak cash flow management, an indication that would suggest that investors should sell the stock if they do not see signs that the problem is likely to be fixed.

TABLE 10-4. REVENUE ACCOUNTING POLICY AND LAWSUITS FOR WEB CONTENT FIRMS

Company	Revenue Accounting Policy	Legal Proceedings
CNET	Advertising revenues are recognized in the period in which the advertisements are delivered.	Simon Property Group filed a complaint against mySimon alleging that "mySimon" mark infringes the Simon Property Group's "SIMON" mark.
Forrester Research	Research revenues are recognized pro rata on a monthly basis over the term of the contract. Its advisory services clients purchase such services with memberships to its research. Billings attributable to advisory services are initially recorded as deferred revenue and recognized as revenue when performed.	None
Giga Information	Amounts received in advance of services provided are reflected in Giga's financial statements as deferred revenues and are recognized monthly on a pro rata basis over the term of the contract.	Giga Media filed a trade name infringement lawsuit.
Jupiter Communications	Research Services contracts are renewable contracts, typically annual, and payable in advance. A portion of billings is initially recorded as deferred revenue and amortized into revenue over the term of the contract.	None
Gartner Group	Revenue from research products is recognized as products are delivered and as Gartner's obligation to the client is completed over the contract period.	Gartner is involved in legal proceedings and litigation arising in the ordinary course of business.
Meta Group	Billings attributable to Meta's Research and Advisory	Meta is a party to certain legal proceedings arising

Company	Revenue Accounting Policy	Legal Proceedings
	Services are initially recorded as deferred revenues and then recognized pro rata over the contract term.	in the ordinary course of business.
CBS MarketWatch	Advertising revenue, derived from the sale of banner advertisements and sponsorships on the company's Web sites, is recognized using the lesser of the ratio of impressions delivered over total guaranteed impressions or on a straight-line basis over the term of the contract in the period the advertising is displayed.	None
TheStreet.com	Revenue from subscriptions is recognized ratably over the subscription period. Deferred revenue relates to subscription fees for which amounts have been collected but for which revenue has not been recognized.	None

Source: Company 10Ks

In light of the rapid rate at which the competitive environment is changing, the ability of managers to adapt to change effectively is a crucial factor in determining the long-term value of a Web content company. To analyze this factor, we can examine the variance of the percentage change in quarterly earnings. Simply put, if a Web content firm is able to increase profits fairly consistently in each quarter, its management is good at adapting to change. Conversely, if a firm's quarterly performance jumps around significantly, its management team is probably less effective at adapting to change. The important point here is that the ability to adapt effectively to change in a rapidly changing environment can be measured by management's ability to produce fairly consistent financial results.

In the case of Web content firms, it is clear that this measure is a somewhat useful one. As Figure 10-4 indicates, consistent earnings growth is particularly important to investors in profitable Web content firms. One of the top-performing firms in the stock market, Forrester Research, generated particularly consistent earnings growth. CNET, whose stock was a leading performer in 1999, actually suffered in 2000 because of its earnings inconsistency after it began its $100 million advertising campaign. This inconsistent performance helps explain the change in CNET's management structure in 2000.

Figure 10-4 also indicates that this low variance is of little predictive value for unprofitable firms. For example, TheStreet.com and CBS MarketWatch had low variances in quarterly earnings change because they were losing money at fairly consistent rates. Not surprisingly, investors do not reward consistently high rates of quarterly increases in losses.

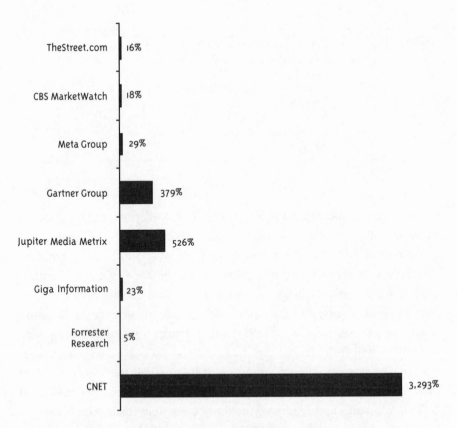

Figure 10-4. Variance in Quarterly Earnings Percent Change for
Selected Web Content Firms, 3/99–3/00

Source: MSN MoneyCentral, Peter S. Cohan & Associates Analysis

In most cases, management integrity and adaptability are not critical factors in differentiating Web content firms from their peers. The average Web content company analyzed here has high levels of management integrity and fairly inconsistent changes in quarterly earnings performance. Forrester Research seems to stand out from its peers in its ability to generate consistent quarter-to-quarter earnings growth. Whether this unique performance makes Forrester Research a potentially good investment depends on further analysis.

Brand Family

Since it is often difficult for investors to understand the details of the Web content process, investors can analyze the firms' brand families as a gauge of the firms' relative value. The conclusion of the analysis of the different firms' brand families is that there are some important differences among the firms, particularly in terms of their customers and principal investors. These differences foreshadow different capabilities of shareholder-value generation and are therefore useful indicators for investors.

The brand families of Web content companies vary depending on whether they are Web-only firms like TheStreet.com or research firms. As Table 10-5 indicates, with the exception of Jupiter Media Metrix, the research firms are not promoting their brand by listing their clients. Traditional research firms—Forrester et al.—seem content to list the number of clients but are reluctant to provide details about the names of their clients. The traditional research firms seem to think that the reputation of the firm should exceed that of its clients. Or the firms may argue that their clients wish to protect their confidentiality. By contrast, the Web-only firms define their brands through their clients, their content partnerships, and the publications where their writers were previously employed. For example, TheStreet.com touts its partnerships with Yahoo, CNET promotes its distribution arrangement with CNBC, and CBS MarketWatch emphasizes its relationships with online trading firms.

TABLE 10-5. MAJOR CUSTOMERS AND PARTNERS OF SELECTED WEB CONTENT COMPANIES

Company	Major Customers/Partners
CNET	USA Networks accounted for 10% of CNET's revenues. Partnerships include CNBC content licensing deal, stakes in Vignette, and beyond.com.
Forrester Research	No data.
Giga Information	1,100 client organizations.
Jupiter Media Metrix	Clients included 19% of the Fortune 500 companies, 15% of the BusinessWeek Global 1000, and 66% of the Media Metrix Top 50 Web and Digital Media properties. Specific clients include Cross Pen, Bear Stearns, America Online, Cablevision, DoubleClick, Hallmark, DLJ, Inktomi, Sony, Goldman Sachs, MediaOne, Putnam Investments.
Gartner Group	9,600 worldwide clients.
Meta Group	5,000 subscribers to its Research and Advisory Services in 2,000 client organizations. Parnership with Computerwire.
CBS MarketWatch	Partner relationships with Quicken, Wingspan Bank, CarClub.com, and E-Loan; joint venture agreement with The Financial Times Group, online brokerage services (Ameritrade, Datek, Salomon Smith Barney, Merrill Lynch, Fidelity, E*Trade, First Trade, Mr. Stock, Multex, and ScoTTrade); co-branded pages with financial organizations, such as The Wall Street Journal Interactive, E*Trade, Fidelity, Waterhouse, Ameritrade, Datek, Merrill Lynch Online, The Street.com, and Charles Scwhab. advertise on a Yahoo, Lycos, Excite, and AltaVista.
TheStreet.com	Content syndication agreements with Yahoo, America Online, MSN Money Central, Salon.com. Advertisers include Ameritrade, National Discount Brokers, Datek, DLJdirect, E*Trade, Compaq, and Exxon/Mobil.

Source: Company Financial Statements (10Ks)

Investors do not appear to prefer the stock of Web content firms based on brand families. CBS MarketWatch and TheStreet.com have formed partnerships with many well-known names. It appears that investors do not assign much value to these partnerships, since the stocks of these firms performed the worst of any of the Web content firms. Most likely, investors recognize that these particular firms have not identified how to convert these brands into profitability. As a result, investors believe that these firms are spending money to associate themselves with well-known names. However, investors have concluded that the costs of these associations cannot generate sufficient revenues to offset the costs. In general, investors do not seem persuaded that branding for Web content firms is an investment; rather, they view branding as an expense. So firms that can get people to pay for their content—such as the research firms—seem to be in a better position to attract investor interest.

A similar conclusion can be reached about the investor reaction to the branding power of major investors in Web content companies. Simply put, investors do not think that power is significant. As indicated in Table 10-6, CNET and Forrester Research, two of the top performing firms, are mostly owned by their founders. Since their founders are not as famous to the general public as some of the investors (e.g., The New York Times Company) in a poor stock market performer such as TheStreet.com, we may conclude that the these names do not bring added luster to the companies.

Nor does ownership by the founders seem to add investment luster. For example, while CNET and Forrester Research are owned primarily by founders, so is Giga Information, which performed relatively poorly in the stock market until its change of management in 2000. The basic conclusion we reached regarding brand family seems to apply both to investors, customers, and partners. Investors in Web content companies do not seem to use brand family as a proxy for future profit potential. Instead, investors pierce the brand veil to analyze financial effectiveness—a test we will examine in the next section of this chapter.

TABLE 10-6. MAJOR INVESTORS IN SELECTED WEB CONTENT COMPANIES

Company	Major Investors
CNET	CNET's investors include chairman Halsey Minor (11%) and CEO Shelby Bonnie (12%).
Forrester Research	Chairman and CEO George Forrester Colony, who founded the company in 1983, owns 65% of Forrester.
Giga Information	Founder Gideon Gartner (founder of IT consulting firm Gartner Group) owns nearly 23%.

Company	Major Investors
Jupiter Media Metrix	Gartner Group (28%), Gene DeRose (chairman and CEO, about 16%), and Kurt Abrahamson (president and COO, about 8%). Joshua Harris, who founded the company in 1986 (but left to launch Internet TV network Pseudo Programs), owns 7.5% of Jupiter Communications.
Gartner Group	IMS Health, which owned just less than half of Gartner Group, distributed its interest in the company to IMS shareholders in 1999. Investment firm Silver Lake Partners is buying about 25%.
Meta Group	Dale Kutnick (14.8%), T. Rowe Price Associates, (13.9%), Marc Butlein (9.4%), Massachusetts Financial Services Company (7.1%), FMR Corporation (5.6%), George McNamee (1.6%).
CBS MarketWatch	Media firm Viacom (owner of CBS) and market data provider Data Broadcasting Corporation (which is majority-owned by UK media giant Pearson) each own about 32% of the company.
TheStreet.com	James J. Cramer (13.0%), Martin Peretz (11.8%), Fred Wilson (1.5%), Jerry Colonna (1.3%), Edward F. Glassmeyer (6.6%), Oak Investment Partners VIII (6.6%), Chase Venture Capital Associates (5.8%), Softbank (5.7%), The New York Times Company (6.2%).

Source: Company Proxy Statements

Financial Effectiveness

Investors use financial effectiveness as an important factor in deciding which Web content firms' stocks to purchase. This distinction comes into focus once we analyze the financial effectiveness of two groups of Web content firms—the Web-only and the Internet research firms. Figure 10-5 helps illustrate this point.

The top performing Web-only firm in the stock market had the highest sales productivity. More specifically, CNET, whose stock price increased the most in 1999, had the highest sales productivity of its peers at $204,000. Conversely, TheStreet.com, with relatively low sales productivity of $71,000, was among the worst-performing Web content stocks.

There is a rougher correlation between sales productivity and stock market performance of research firms. Forrester Research, with sales pro-

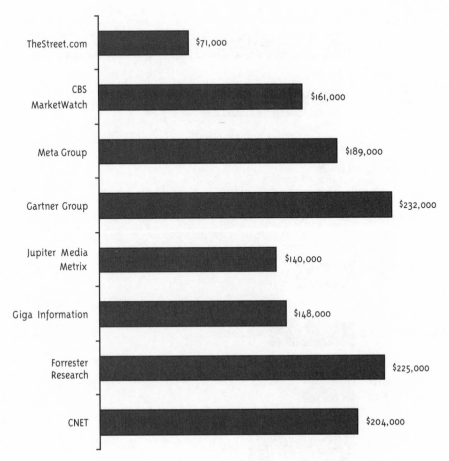

Figure 10-5. Sales Per Employee of Selected Web Content Firms, 2000
Source: MSN MoneyCentral

ductivity of $225,000 per employee, had the best stock market perform-
ance of any research firm. Gartner Group's sales productivity was slightly
higher, but its stock market performance was far weaker. Here it is likely
that Gartner's underperformance in the stock market was caused by some
of the management weaknesses to which we referred earlier. It is entirely
possible that the decision by Silver Lake Partners to invest in Gartner
Group may have been prompted by their belief that good management
could revive Gartner Group's sales and profit growth.

The other measure of financial effectiveness, the SG&A-to-sales ratio,
tells a similar story. As Figure 10-6 indicates, the two worst-performing
Web content stocks had the highest SG&A-to-sales ratios. More specifically, CBS MarketWatch and TheStreet.com had SG&A-to-sales ratios of

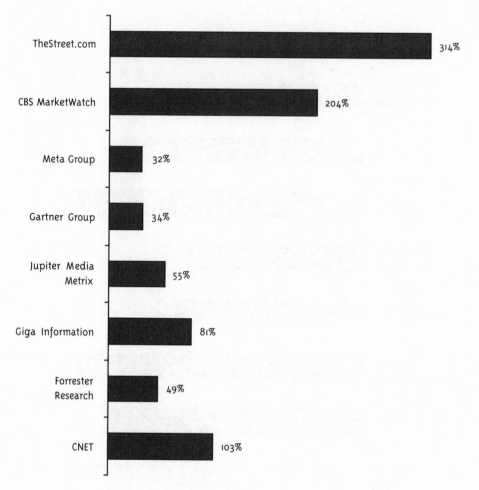

Figure 10-6. Selling, General, and Administrative Expense as a Percent
of Sales for Selected Web Content Firms, 1999
Source: MSN MoneyCentral

204% and 314%, respectively. Clearly, both firms think of themselves as making important investments in the future value of their assets. Investors, as noted earlier, tend to view SG&A as expenses, not investments. Or perhaps more precisely, investors have a difficult time estimating the cash flows that are likely to be generated by spending this money.

Among the research firms, however, it appears that, within a band, investors reward firms for spending more on SG&A. More specifically, firms spending between 45% and 50% of SG&A to sales seem to do better in the stock market than firms such as Meta and Gartner Group, which spend below 35%, and Giga Information, which spent 81%.

While it would probably be a mistake to conclude that a correlation between stock price performance and these specific levels of SG&A to sales will be maintained in the future, it is likely that investors will continue to reward research firms that spend an optimal share of their sales on SG&A. Simply put, investors seem to have concluded that within a certain range, SG&A is an investment for research firms that is likely to yield a return. Above a certain level, investors perceive a firm as wasteful. Below a certain level, investors perceive that the firm is not investing sufficiently.

Relative Market Valuation

The stock market's approach to valuing Web content companies suggests that firms in the industry appear fairly valued relative to each other. As Table 10-7 indicates, on the basis of price/sales ratio divided by revenue growth rate, the market is assigning valuations that appear to reflect the relative positions of the firms in the industry.

TABLE 10-7. STOCK PRICE AND REVENUE GROWTH PERCENT CHANGE, 2000, AND VALUATION RATIOS FOR SELECTED WEB CONTENT FIRMS, JANUARY 12, 2001

Company	Stock % Change	Revenue Growth	Price/Sales	(Price/Sales) /Rev. Growth	Price/Book
CNET	−72%	107%	1.16	1.08	0.41
Forrester Research	45	83	8.72	10.51	6.76
Giga Information	18	39	0.40	1.03	negative
Jupiter Media Metrix	−74	202	3.32	1.64	0.45
Gartner Group	−55	17	0.82	4.82	9.05
Meta Group	−66	26	0.47	1.81	0.47
CBS Market Watch	−68	162	1.16	0.72	0.41
TheStreet.com	−92	107	2.65	2.48	0.81
MEAN	**−47%**	**93%**	**2.34**	**3.01**	**2.62**

Source: MSN MoneyCentral

The firm whose stock prices has appreciated the most has a double-digit ratio of (price/sales)/revenue growth. For example, Forrester Research's is 10.51. With the possible exception of Gartner Group and TheStreet.com, the valuations of the firms seem to line up with their relative stock market performance. The latter two firms appear to be somewhat overvalued, given their relatively slow rate of revenue growth. It is possible that Gartner Group is perceived as being more valuable because of the investment from Silver Lake Partners, whose reputation commands some respect among institutional investors. If their judgment proves prescient, then this potentially high valuation may prove justified. Of course, it remains to be seen whether Gartner Group will be able to accelerate its revenue and profit growth. Until then, its stock simply appears to be somewhat overvalued.

The Internet Investment Dashboard analysis suggests that Web content is not likely to be an attractive area for investors, although this could evolve. In Figure 10-7, three of the factors are hatched and three are dotted. The hatched indicators are the economic bargaining power, closed-loop solution, and financial effectiveness. The reason for this, as we have seen, is that the average Web content firm lacks sufficient economic bargaining power to negotiate a price high enough to offset its costs; it offers

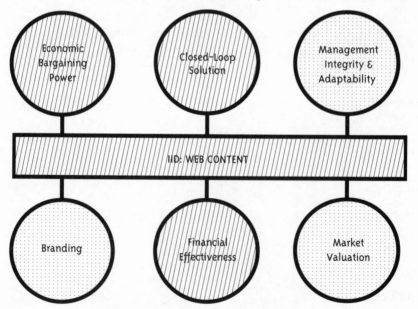

Figure 10-7. Internet Investment Dashboard Assessment of Web Content Segment
Key: Dotted = mixed performance and prospects; hatched = poor performance and prospects.

a closed-loop solution to visitors but not to advertisers, and it does not make the most effective use of its capital.

Offsetting these negatives somewhat, the average Web content firm has high management integrity, although its ability to adapt effectively to change is not consistent. Furthermore, many firms have compelling brand families, while others do not. Finally, the market valuations of many firms in the Web content industry are aligned with their relative performance on the other indicators. These indicators suggest that investors should approach the Web content sector with caution. The source of revenue for the industry is rather ill-defined but is expected to consist of subscriptions to research services and advertising. The economic bargaining power of most participants is so low that the average profit margin is −62%.

Web content is an industry where providing a closed-loop solution is particularly important. Few firms actually provide a closed-loop solution to the two groups of stakeholders on whom they rely. For example, while many firms provide a closed-loop solution for those seeking information, they provide a less compelling solution for advertisers.

Web content firms have relatively high levels of management integrity, although their ability to adapt to change is mixed. The brand family of many Web content firms is strong, while some others have weak brand families. The financial effectiveness of Web content firms is generally not strong, and the relative market valuation of many firms seems to make sense, with a few exceptions.

PRINCIPLES FOR INVESTING IN WEB CONTENT STOCKS

The analysis we have conducted here is subject to change. Some of the strong-performing companies will falter, some of the weak ones could generate positive surprises, and new competitors could emerge to tilt the playing field in a totally different direction. As a result, investors in Web content stocks need some principles to guide them through the analytical process as investment conditions evolve.

Web Content Investment Principle 1. Analyze the effectiveness of the strategies of Web content firms such as CNET to extend their services to the mass market. The most compelling potential investment opportunity in the Web content field could emerge if CNET can apply to the general public the principles that worked for technology-focused users of its content—using mySimon as an enabler. If CNET succeeds in its quest to create a mass-market brand, it could generate much higher sales and profit growth. Investors should watch carefully to see whether there are early signs of such success. If CNET succeeds in this effort, it could be an attractive investment.

Web Content Investment Principle 2. Assess whether firms such as Gartner Group and Giga Information are able to create effective closed-loop solutions specifically focused on Web content. While most Web content firms have not created closed-loop solutions for advertisers, many have done so for the consumers of content. Gartner Group and Giga Information appear to be undergoing turnaround efforts that could result in closed-loop solutions for both Internet technology companies and customers of Internet technologies. If such a solution is actually created, these companies could prove to be significantly undervalued. Investors should monitor the progress of these companies' turnaround efforts. If they show signs of succeeding, the stocks could be good investments.

Web Content Investment Principle 3. Avoid investing in firms where there are questions about management adaptability. Our analysis indicated questions about management adaptability in a few firms. Clearly, most firms in the Web content industry have not come close to implementing a business model that offers hope of generating a profit. CNET comes closest and is most likely to succeed in building a profitable business model in the future. The Internet research firms could become more profitable if they can accelerate their revenue growth. The other firms may not be able to generate profits in the future unless they can develop a fairly original and effective new source of revenue.

Web Content Investment Principle 4. Assess the extent to which Web content firms are able to monetize their brands. As we noted several times in this chapter, investors do not give Web content firms much credit for assembling compelling customers and partners unless they convert those relationships into significant revenues and profits. If Web content firms begin to find ways to achieve this monetization, then investors should take a harder look at investing in the companies that succeed.

Web Content Investment Principle 5. Use financial effectiveness measures to distinguish among long-term leaders, medium-term survivors, and liquidation candidates. While investors seem to have tolerated the absence of financial discipline among Web content firms up until April 2000, this patience is fading. Firms that lack financial discipline are unlikely to survive as independent entities unless they can obtain outside financing, and this has become more difficult. Firms that generate strong profitability and rapid revenue growth are likely to be rewarded with more highly valued stock, which they can use as currency to acquire valuable assets. Such currency can contribute to a virtuous cycle of profitable growth if it is used wisely. Investors should seek out Web content firms whose financial effectiveness is improving and avoid those whose financial effectiveness is eroding.

Web Content Investment Principle 6. Monitor relative valuations among Web content sectors and within Web content sectors to identify potentially under- and overvalued companies. As we noted earlier, investors seem to have assigned fairly reasonable valuations to most of the Web content firms. As a result, currently there are not many candidates for closer evaluation to assess whether they might be under- or overvalued. As we noted, investors should pay particularly close attention to Gartner Group. Silver Lake Partners' decision to invest in the firm could signal hidden value—or it could be a mistake.

CONCLUSION

Web content is generally an unattractive area for investors. While one or two firms have the potential to break out of the doldrums, such a breakout depends on their ability to execute fairly fundamental changes in corporate direction. While the concepts for the new strategies may make sense, investors must monitor closely whether these concepts can be translated into significantly improved financial results. Until they can, Web content remains an area that investors should avoid unless they are particularly prone to investing in long shots.

Chapter 11

TAKING A FREE RIDE: INTERNET SERVICE PROVIDER STOCK PERFORMANCE DRIVERS

INTERNET SERVICE PROVIDERS (ISPs) were among the first segments in which private investors made serious money. The idea that ISPs would be the choke point for the tremendous growth potential of the information superhighway was quite a powerful one in the mid-1990s. Since then, however, reality has sunk in. There are virtually no independent, publicly traded ISPs left because the proliferation of competitors knocked pricing down to such a low level that the price of a ubiquitous, always-on connection is very inexpensive—far lower than the cost of operating and marketing the ISP. There are no attractive ISP investments left. Those with the potential to be attractive, such as AOL, are no longer really ISPs—they have evolved into something more.

Exodus Communications is among the best ISP stock market performers that remain independent. Exodus's stock price increased 180%, from $20 in September 1999 to $56 in September 2000. The news on Exodus Communications is not all good, however. The September 2000 price represents a 37% decline from the company's 52-week high of $89, reached in March 2000; and by December 2000 it had dropped back to $20. As we shall see later in this chapter, Exodus's Web hosting service enables dot-coms and land-based competitors to outsource the operations of their Web sites. Such outsourcing has enabled Exodus to capture a rapidly growing share of the revenues associated with e-commerce. The 78%

decline in Exodus's stock price since the April 2000 Internet stock crash reflects investors' skepticism about the firm's ability to sustain precrash growth rates. Some customers may not be in a strong position to pay their Web hosting bills, and others may choose to cut back on the pace of their e-commerce operations.

EarthLink is a compelling example of the profound weakness of the B2C ISP business model. EarthLink's stock price declined 83%, from $30 to $5, from September 1999 to December 2000. After having acquired its competitor MindSpring, EarthLink has seen its stock price decline fairly steadily as it becomes increasingly difficult for investors to understand how firms in the consumer ISP segment will be able to generate profits in the future—particularly in light of competition from free ISPs. With huge operating losses and brutal price competition, it is hard to see a way out for shareholders. It is difficult to imagine an acquirer paying much of a premium for a company with EarthLink's financial statements.

When the Internet became a significant commercial phenomenon in the mid-1990s, some investors made huge gains by investing in companies that could connect individuals and businesses to the Internet. One example is UUNET Technologies, which was acquired by WorldCom. This success helped encourage about 6,500 companies to jump into the ISP market. These ISPs range from local firms whose territory covers a few cities to global powerhouses like AOL. As we will see, AOL is a case unto itself, with many sources of revenue extending far beyond Internet access provision.

From an investor's standpoint, ISPs fall into two categories: those focused on providing businesses with access to the Internet (B2B ISPs) and those focused on consumers (B2C ISPs). The B2B ISPs are growing rapidly and appear to have the advantage of playing an important role in the survival of their clients' e-business operations. As a result, B2B ISPs have some economic bargaining power with customers. Given the large number of ISPs in general, this bargaining power is not unlimited. B2C ISPs seem to have even less economic bargaining power, since the presence of free ISPs such as Juno Online and others eliminates the pricing power of these B2C ISPs. As we will soon see, these differences are reflected in the different rates of stock price appreciation among the two groups of ISPs.

ISP STOCK PERFORMANCE

We begin with an examination of the stock price performance of the ISP segment. We noted in Chapter 3 that the stock market performance of a cross section of ISP firms was 366% in 1999. This average performance compares favorably to the NASDAQ's 86% increase in 1999 and to the 339% average stock price increase of the companies in our nine-segment Internet stock index in 1999. In 2000, ISPs lost 69% of their value, worse than the 67% drop in the Internet average.

One important factor behind 1999's strong ISP stock performance, as illustrated in Figure 11-1, was the explosive stock price appreciation of two publicly traded ISPs, Metricom and Exodus Communications. Metricom, whose stock price increased 1,416% in 1999, benefited from two powerful forces. The first was the market's positive reaction to Microsoft cofounder Paul Allen's decision to purchase over 30% of Metricom's stock. Since many investors believe that Allen is a genius, they concluded that his decision to purchase the stock indicated that it had hidden value. The second phenomenon that drove the decision was Metricom's efforts to build a wireless ISP. This is important because wireless Internet access is currently considered to be "hot." As we will see, both of these forces did not translate into excellent financial performance.

Exodus Communications, whose stock price increased 1,006% in 1999, has benefited from explosive growth in demand for its services. Many investors perceive Exodus as the leading B2B ISP, and Exodus has benefited from the management of Ellen Hancock, a former IBM executive. Without the performance of Exodus and Metricom, the publicly traded ISPs would not have fared as well.

The effect of the April 2000 crash in Internet stocks affected different ISPs differently. The market did not uniformly destroy the value of all ISPs as it did the value of the e-commerce stocks. As Figure 11-2 (page 280) indicates, some ISP firms saw their stock prices deteriorate far more than others. On the other hand, one ISP actually experienced an increase in value in 2000.

Many of the ISPs whose stock prices were punished most severely were those that had big operating losses. For example, the B2C ISPs—EarthLink and Excite@Home—both suffered from investors who abandoned these companies out of fear for their continued ability to survive in the event that the public markets no longer cared to invest further—

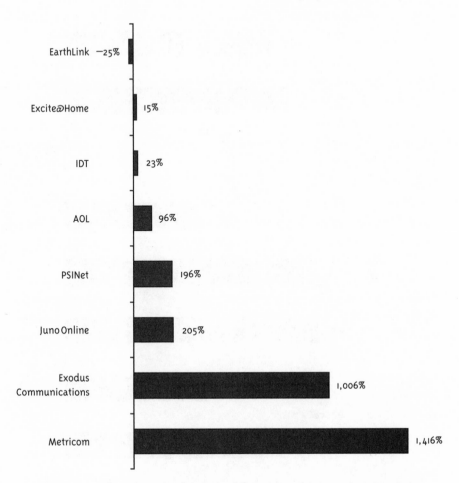

Figure 11-1. Selected ISP Firms' Percent Change in Stock Price, 12/31/98–12/31/99
Source: MSN MoneyCentral

thereby leading to severe cash flow problems. Metricom's reversal of fortune no doubt played into investors' decision to sell that stock. The market also seemed to have a limited regard for AOL's January 2000 announcement of its acquisition of Time Warner—knocking about 54% off its value.

By contrast, a couple of profitable ISPs actually saw an increase in their share prices during the first six months of 2000. Some of the biggest beneficiaries of investor interest were ISPs involved in acquisitions. For example, Verio announced that Japanese telecommunications leader NTT would acquire the remaining Verio shares that it did not already own, driving a 24% increase in the stock's price. Concentric Networks announced that it would be acquired by NextLink. Finally, IDT benefited from an increase in value of its holdings in Internet telephony firm Net2Phone.

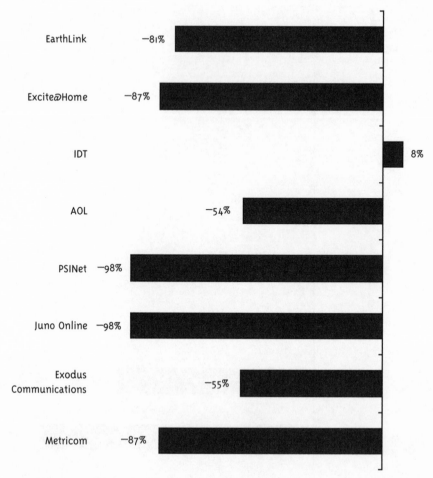

Figure 11-2. Selected ISPs' Percent Change in Stock Price, 12/31/99–12/29/00
Source MSN MoneyCentral

INTERNET INVESTMENT DASHBOARD ANALYSIS OF ISPs

While the foregoing discussion of the relative stock price performance of ISP segments is suggestive, it does not offer a fully satisfactory explanation. To obtain such an explanation, we apply the Internet Investment Dashboard (IID) analysis to the ISP segment.

As we shall see, the IID analysis suggests that the ISP segment is only likely to generate attractive investor returns in the future for a small number of investments. The ISP industry clearly lacks economic bargaining power, as its low entry barriers have admitted thousands of competitors. While many providers offer customers a closed-loop solution, many of these providers continue to have trouble getting compensated for the provision of that solution. Many industry participants have demonstrated their ability to adapt effectively to change and have created a compelling brand family, and these factors help explain their superior performance in the stock market. Furthermore, a few firms have managed their financial affairs quite effectively, and some of these companies' shares are relatively inexpensive. As a result, investors may find a few good opportunities within the ISP segment.

Economic Bargaining Power

The ISP market consists of two significant segments, neither of which has sufficient economic bargaining power to be profitable for the average participant. One segment provides access to the Internet; the other hosts Web sites for organizations and individuals. As Figure 11-3 indicates, according to International Data Corporation (IDC) total ISP revenues in the U.S. are expected to grow at a 54% compound annual rate, reaching $66.8 billion in 2003. IDC predicts that dedicated connections to the Internet for small and medium-sized businesses will grow from approximately 90,000 in 1996 to just under 800,000 in 2000, representing a 73% compound annual growth rate (PSINet 1999 Annual Report).

The growth of the Internet has resulted in a highly fragmented industry of 6,700 national and local ISPs in the United States, with no dominant ISP serving the needs of small and medium-sized businesses. Independent regional and local ISPs have successfully captured approximately one-half of this market, despite the substantially greater resources of the national providers. However, rising costs and increasing demands from business customers have made it more difficult for the small ISP to meet its customer's demands on a cost-effective basis.

In addition, as more businesses move from establishing an Internet presence to using the Web for secure connections between geographically dispersed locations, remote access to corporate networks, and B2B e-commerce, the demand for Internet connectivity and related services is expected to grow. As illustrated in Figure 11-4, the Gartner Group forecasts that World Wide Web hosting services revenue will grow from $896 million in 1998 to $5.9 billion in 2003, a compound annual rate of 46% (PSINet 1999 Annual Report).

Despite the rapid growth and substantial size that has been forecast for the ISP and Web hosting markets, the average ISP is a money-loser. More

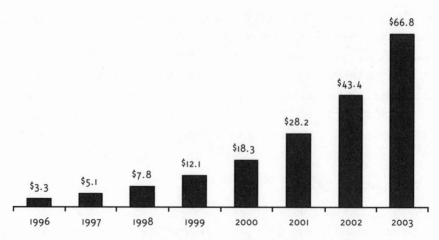

Figure 11-3. ISP Revenues, 1996–2003, in Billions of Dollars
Source: International Data Corp.

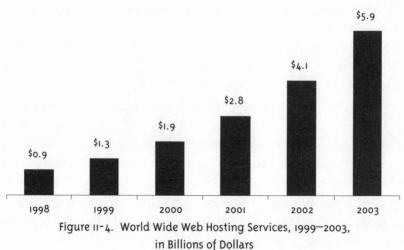

Figure 11-4. World Wide Web Hosting Services, 1999–2003,
in Billions of Dollars
Source: Gartner Group

specifically, as Table 11-1 illustrates, the average ISP in the sample analyzed here generates net losses of 98 cents for every dollar of revenue. In
fact, however, this average performance masks some significant variations
among various members of the group. For example, Excite@Home's net
margin is an appalling –448%, while AOL's net margin is a comfortable
19%. While Excite@Home suffers from an awkward ownership structure,
a lagging market share, and an unfocused business strategy, AOL is firing

on all cylinders—generating revenues from monthly connection charges, e-commerce, advertising, and partnership deals.

TABLE 11-1. MARGINS FOR EIGHT ISPS, DECEMBER 2000

Company	Stock % Change	Gross Margin	Operating Margin	Net Margin	2-Year Trend
Metricom	−87%	−805%	−1,064%	−1,064%	—
Exodus	−55	59	−37	−37	+
Juno Online	−98	20	−145	−145	+
PSINet	−98	33	−72	−72	—
America Online	−54	55	30	19	+
IDT	8	20	41	21	+
Excite@Home	−87	85	−448	−448	+
EarthLink	−81	48	−26	−26	+
MEAN	**−69%**	**46%***	**−94%***	**−98%***	**0**

Source: MSN MoneyCentral

Note: Two-year trends represent the extent to which net margins have changed in the most recent two years. + indicates a small improvement, 0 indicates no change, —indicates large deterioration.

*Excludes Metricom

A comparison of the relative economic bargaining power of Excite@Home and AOL can provide investors with useful insights for evaluating future investment opportunities in the ISP industry. The basic concept behind these two B2C ISPs is to build a media company with so many registered users that it becomes a magnet for organizations trying to sell products and services to those registered users. Learning lessons from traditional media, Excite@Home and AOL both tried to obtain as many users as possible. As we will see, AOL has managed to maintain the number-one position in terms of relative market share, while Excite@Home's position has lagged considerably despite its efforts to increase relative market share through acquisitions of sites such as the online greeting card service provider bluemountain.com.

Relative market share, as measured by firms such as Jupiter Media Metrix, is the most powerful factor determining an ISP's economic bargaining power. AOL, which has by far the leading market share, is in a very strong position to negotiate with many different service and technology providers. This negotiating position enables AOL to extract significant amounts of cash from partners in exchange for letting these partners get some form of access to its paying users. By contrast, Excite@Home has far fewer users and is therefore not in a strong bargaining position with potential advertisers and e-commerce partners.

Unfortunately for investors, it is difficult to imagine any imminent change in relative market share that some smarter investors could take advantage of. One clear lesson from this analysis is that investors may not get significant price appreciation if they invest in ISPs that are not market leaders—unless there is a likelihood that the firm will be acquired for an attractive price. Another important conclusion from the analysis of ISP economic bargaining power is that rapid growth in a market does not imply profitability—nor does it imply that investors should pile into stocks in that sector without careful analysis.

Closed-Loop Solution

The notion of a closed-loop solution is implemented differently for B2C and B2B ISPs. For B2C ISPs, a closed-loop solution requires providing consumers with access to the Internet as well as offering information services, e-mail accounts, real-time messaging, and e-commerce. B2C ISPs also must offer closed-loop solutions for advertisers, which involves the ability to direct information about advertisers' products to specific groups of customers who are more likely to purchase those products.

B2B ISPs face a different set of challenges in offering their customers closed-loop solutions. The B2B ISP must provide an organization with secure Internet access and with good backup facilities, in addition to having the ability to add capacity as clients' demands expand. In the case of Web hosting, a closed-loop solution also includes the provision of dedicated servers with software to support the Web site operations.

One way to assess the relative effectiveness of the strategies of ISPs is to analyze the firms' relative size, revenue growth, and market capitalization. This analysis can help us gain insights into which firms are generating the greatest product market momentum and the extent to which that momentum has translated into greater stock price appreciation.

The correlation between momentum in the product markets and in the capital markets is partially borne out by the evidence. Comparing Tables 11-1 and 11-2, it appears that investors are distinguishing between ISPs seemingly heading for extinction, such as Metricom and Juno Online, and ISPs likely to survive. ISPs that are likely to survive enjoy market capitalization in excess of $1 billion. Those likely to fail are losing huge amounts of money (net margins worse than –100%). By contrast, ISPs that are profitable or not losing too much money, such as AOL and Exodus, enjoy substantial market capitalization.

TABLE 11-2. REVENUE MOMENTUM OF EIGHT ISPS, JANUARY 12, 2001, IN MILLIONS OF DOLLARS

Company	Stock % Change	12-Months Revenue	Revenue Growth	Market Capitalization
Metricom	–87%	14	–30	417
Exodus	–55	646	287	9,080
Juno Online	–98	102	148	65
PSINet	–98	1,040	128	454
America Online	–54	7,394	35	198,300
IDT	8	1,094	49	771
Excite@Home	–87	528	200	3,520
EarthLink	–81	1,134	197	1,030

Source: MSN MoneyCentral

Generally, the B2C ISPs saw their stock prices decline more than the B2B ISPs. The best-performing B2C ISP in the stock market, AOL, is not really a pure ISP. As noted earlier, AOL is a combination of a Web portal, ISP, and e-commerce company. AOL's stock price appreciated a respectable 96% in 1999, dropping back 53% in 2000 after its announcement of the Time Warner acquisition. AOL, while much larger in terms of revenues and market capitalization than all of its peers combined, posted relatively slow revenue growth of 35%. It remains to be seen whether this revenue growth will decline now that the merger is complete.

Excite@Home and EarthLink, while much larger in revenue and much faster-growing than their B2B peers, did not experience stock price appreciation of anywhere near the level of these smaller peers. More specifically, Excite@Home and EarthLink increased revenues at 200% and 197% respectively, while their stock prices declined 87% and 81% respectively in 2001.

Behind the numbers there were significant differences in the strategies of the ISPs that drove their relative stock market performance. For example, Exodus Communications was the clear leader in Web hosting for organizations. Because Exodus focused exclusively on building a Web hosting capability that actually satisfied organizations' need for a robust, secure site that could expand as customers' needs changed, it experienced the most rapid revenue growth and did very well in the stock market. Verio pursued a strategy of acquiring small and medium-sized ISPs. While this strategy contributed to growth, it may have also created challenges of integrating the acquired operations, which contributed to NTT's decision to acquire Verio in 2000.

The relative stock market performance of the B2C ISPs also reflects the relative effectiveness of their strategies to bind consumers and advertisers to the sites. For example, AOL has done an excellent job of attracting and retaining individual subscribers through skillful consumer marketing, reaching a base of over 27 million subscribers. This leading audience share has made it easier for AOL to charge advertisers and e-commerce partners substantial fees for enabling them to gain access to that audience.

Excite@Home, by contrast, has been faced with many complex internal challenges that make it difficult to offer a truly closed-loop solution. For example, Excite@Home's significant ownership by AT&T can often slow down the pace of decision-making required to install sufficient cable capacity to meet demand and invest in the service operation needed to help those customers with technical problems. Excite@Home also faces the challenge of integrating its content with its cable service and generating a sufficiently large audience to compete effectively with AOL in generating advertisers. Clearly, investors have lost patience with Excite@Home's efforts to compete and have driven down its shares accordingly.

A more in-depth analysis of the strengths and weaknesses of the ISPs profiled here reveals not only the relative market positions of the firms but also the reasons behind those relative positions. Table 11-3 summarizes the key strategic differences among selected firms profiled in this chapter. For example, an analysis of Metricom's strategy suggests that the relatively small number of cities in which the service operates has the potential to limit the revenues that it can generate. Furthermore, the costs of building the infrastructure required to make Metricom's service a more national one could be higher than the company can finance.

The table also highlights the huge gap in market share that B2C ISPs must close in order to catch up with AOL. For example, AOL has 25 million subscribers, which is almost 10 times more than the number-two ISP, EarthLink, which has about 3 million subscribers following its acquisition of MindSpring. While EarthLink has continued to make acquisitions, including the announced purchase of OneMain in June 2000, the market share gap to be closed is formidable. Similarly, Excite@Home, with its 1.2 million subscribers, faces an uphill battle as it tries to deal with complex shareholder arrangements, acquisitions, and frustrated consumers and shareholders.

TABLE 11-3. STRATEGIC DIFFERENCES AMONG EIGHT ISPs

Company	Competitive Strengths/Weaknesses
Metricom	*Strengths:* Ricochet's fixed-wireless network offers subscribers access to the Internet and corporate data networks in San Francisco, Seattle, and Washington, D.C., metro areas; it is also available in 12 U.S. airports; Southern California Edison accounts for 10% of sales.
	Weaknesses: Significant cash flow requirements may exceed financing opportunity; limited geographic scope.
Exodus	*Strengths:* 20 Internet Data Centers where clients store their servers in secure vaults; furnishes maintenance and network connections; expanding its geographic penetration and security services offerings through acquisitions.
	Weaknesses: Unprofitable; significant investment required to sustain growth.
Juno Online	*Strengths:* 3.38 million active subscribers (including 730,000 billable subscribers); 300 advertisers.
	Weaknesses: Too many nonpaying subscribers; too dependent on advertising; high subscriber acquisition costs.
PSINet	*Strengths:* worldwide fiber optic network with 600 points of presence (POPs), serving 22 countries in North and South America, Europe, and the Asia/Pacific region; growth has come through ISP acquisitions.

Company	Competitive Strengths/Weaknesses
	Weaknesses: Weak financial condition; challenges of acquisition integration.
America Online	*Strengths:* 27 million subscribers; leading ISP market share; CompuServe Interactive Services adds 2.5 million additional subscribers; owns Netscape Netcenter portal and software; also owns AOL Instant Messenger, AOL.com portal, Digital City, and AOL MovieFone.
	Weaknesses: Difficult to assess benefits of merger with Time Warner.
IDT	*Strengths:* Callback technology allows overseas phone callers to bypass non-U.S. carriers' high rates; wholesale carrier services to 125 customers, and sells calling cards; network includes 70 switches and 16 undersea fiberoptic cables; provides Internet access to 80,000 customers and 500 businesses; 48%-owned spinoff Net2Phone lets users use Internet for telephone calls.
	Weaknesses: Howard Jonas's ownership is of concern to some investors.
Excite@Home	*Strengths:* High-speed Internet access, content and multimedia applications, to 1.2 million customers; FreeWorld service offers free Internet access.
	Weaknesses: Weak financial condition; ownership by AT&T; small market share.
EarthLink	*Strengths:* #2 Internet service provider in the U.S.; 3 million consumer and business customers, provides dial-up, DSL, single ISDN, and cable modem Internet access; offers domain registration, e-commerce, and Web hosting.
	Weaknesses: Sprint ownership is a mixed blessing; weak financial performance; challenge of merging cultures from acquisitions.

Source: Company Reports, Peter S. Cohan & Associates Analysis

Another important point that investors should note from this analysis of ISP strengths and weaknesses is the predominance of mergers. For example, NTT bought Verio, NextLink purchased Concentric Networks, AOL bought Time Warner, Sprint has the potential to acquire EarthLink, and EarthLink is acquiring OneMain. These examples suggest that investors should not be surprised to see more consolidation in the ISP industry.

Investors need to be able to evaluate which firms are most likely to benefit in terms of stock price appreciation from these acquisitions in the future. Given the strong stock price appreciation of Verio and Concentric Networks after the announcement of their mergers, investors could benefit from anticipating acquisition targets—particularly if a target's stock price has been declining for an extended period of time. Investors must recognize that such evaluations are not simple. For example, EarthLink or Excite@Home might appear on the surface to be candidates for acquisition. An analysis of their ownership structure, however, reveals that an acquirer would need to negotiate with powerful telecommunications firms in order to wrest control of the firms, since these two firms are owned substantially by AT&T and Sprint, respectively. Better potential acquisition candidates might be smaller ISPs, such as OneMain and others, which may have less complex ownership structures.

In conclusion, an ISP's ability to offer customers a closed-loop solution makes a big difference to investors. The relative strength of an ISP's capabilities in creating value for customers tends to determine how rapidly the firm can grow relative to its peers. This growth rate often correlates with the relative stock price performance. Perhaps more importantly, this partial correlation reveals that investors seem to use an analysis of an ISP's relative strengths and weaknesses as an important factor in determining whether or not to invest in the firms' stock.

Management Integrity and Adaptability

Management integrity and adaptability are important considerations for investors in ISPs. While most of the ISPs analyzed in this chapter have strong management integrity, the exceptions seem to be the companies whose stock prices have not done so well. As to adaptability, it appears that many firms have been changing their corporate strategies aggressively; however, some have been more effective than others at achieving the improvements in performance that were hoped for when these strategy changes were conceived. Investors have rewarded the ISPs that have shown high integrity and that have succeeded at improving performance through effective adaptation.

An analysis of legal proceedings and accounting policies for the ISPs covered in this chapter suggests that, with a few exceptions, all ISPs have relatively high levels of management integrity. As Table 11-4 indicates, all the ISPs except IDT, Excite@Home, and AOL have negligible legal proceedings against them, suggesting that they have avoided integrity issues. The exceptions to this general pattern seem to have been hurt by investors. In fact, their relatively weak stock market performance could be partially attributable to the integrity questions raised by the lawsuits. In the case of IDT, for example, there have been claims of employment discrimination and breach of contract with partners. These legal issues are at an inconclusive status as of this writing, yet they suggest potential problems with management integrity. Since IDT's financial performance appears strong, it is possible that investors' concerns with integrity played a role in IDT's relatively weak stock market performance.

TABLE 11-4. REVENUE ACCOUNTING POLICY AND LAWSUITS FOR SELECTED ISPs

Company	Revenue Accounting Policy	Legal Proceedings
Metricom	Product revenues are recognized upon shipment. Service revenues are recognized ratably over the service period.	None
Exodus	Revenues are generally billed and recognized ratably over the term of the contract.	None
Juno Online	Advertising revenues recognized as earned.	Two sexual harassment lawsuits.
PSINet	PSINet recognizes revenue when persuasive evidence of an agreement exists, the terms are fixed, services are performed, and collection is probable.	Relatively immaterial claims of infringement of proprietary rights.
America Online	Subscription services revenues are recognized over the period that services are provided.	Complaints alleging breaches of fiduciary duty in regard to Time Warner merger.
IDT	Telecommunication, Internet telephony service, Internet subscription service, and debit card revenues are recognized as service is provided.	Surfers Unlimited alleges interference with prospective business advantages, breach of contract, and improper use of confiden-

Company	Revenue Accounting Policy	Legal Proceedings
		tial and proprietary information; six former employees alleged employment discrimination.
Excite@Home	Subscription revenue recognized in the period in which subscription services are provided.	Complaints alleging breaches of fiduciary duty to Excite@Home and its public stockholders; complaint alleging violations of the federal antitrust laws regarding exclusive distribution and sales arrangements with cable partners.
EarthLink	Revenues are recorded as earned.	None

Source: Company 10Ks

The potential integrity issues raised by the lawsuits against AOL and Excite@Home appear to be of a different caliber. While in the past AOL was found to be too aggressive in its accounting for marketing expenses, this particular issue seems to have been resolved. The current challenge AOL faces is one of proving that its business judgment in acquiring Time Warner will pay off for shareholders. While the acquisition exposed AOL to a legal complaint, the complaint in and of itself does not suggest significant integrity issues for AOL management.

Excite@Home's legal challenges reflect potentially deeper problems. For example, the merger between Excite and @Home can hardly be said to have produced outstanding financial results. Nor has the stock price performance produced happy shareholders. In fact, the stock price performance has created grounds for shareholder lawsuits against Excite@Home management and directors.

Furthermore, the complaints alleging violation of antitrust laws in regard to Excite@Home's cable-distribution practices certainly put at risk one of the most significant sources of advantage that Excite@Home previously felt it enjoyed—the ability to control information access via cable to the homes to which it is wired. If Excite@Home is ultimately forced to open access to competitors, then the value of the cable investment will be significantly diminished, as will the value of the concept behind the merger between the two firms.

In light of the rapid rate at which the competitive environment changes, the ability of managers to adapt to this change effectively is a crucial factor in determining the long-term value of an ISP. To analyze this factor, we can examine the variance of the percentage change in quarterly earnings. Simply put, if an ISP is able to increase profits fairly consistently in each quarter, its management is good at adapting to change. Conversely, if a firm's quarterly performance jumps around significantly, its management team is probably less effective at adapting to change. The important point here is that the ability to adapt effectively to change in a rapidly changing environment can be measured by management's ability to produce fairly consistent financial results.

In the case of ISP firms, it is clear that this measure is most useful for evaluating the stock market performance of the weakest firms. As Figure 11-5 indicates, the three firms with the highest variances in quarterly earnings changes did the worst in the stock market in 1999. For example, EarthLink—whose earnings-change variance over five quarters was 27,125%—had by far the worst stock market performance. On the other end of the spectrum, with the exception of Metricom, firms with the lowest quarterly earnings-change variance did the best in the stock market. It should be emphasized that these firms were all losing money during the period, so the variance really tells us that the best-performing firms tended to generate fairly consistent changes in losses per share each quarter. Simply put, this analysis is of most use in helping to identify which ISPs investors should avoid owing to their extremely unpredictable financial performance.

As we alluded to earlier, investors ought to judge an ISP management's adaptability based on how well it conceives and executes changes in corporate strategy. AOL and Excite@Home are useful cases to illustrate these differences. AOL has achieved its market dominance not because of its skill at technology but as a result of its ability to create a mass-market consumer brand. This consumer marketing skill began with AOL's blanket-bombing with diskettes and CD-ROMs touting 100 free minutes of service. AOL continued to bolster its consumer mass marketing capability by hiring MTV inventor Bob Pittman as president, followed by skillful acquisitions of ICQ, the instant messaging service,

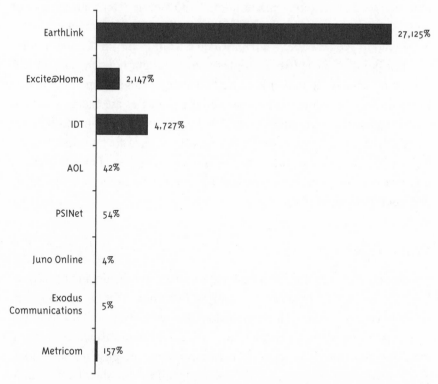

Figure ii-5. Variance in Quarterly Earning's Percent Change for Selected ISPs, 3/99—3/00

Source: MSN MoneyCentral, Peter S. Cohan & Associates Analysis

Compuserve, and Netscape. AOL's skill in acquisitions has been accompanied by significant growth in revenues, revenue streams, profits, and market capitalization.

By contrast, the concept of merging @Home's proprietary cable access to homes with Excite's content has fallen flat on its face. Excite@Home has had significant management problems—resulting in the departure of Excite@Home's CEO Tom Jermoluk to Kleiner Perkins. (Incidentally, John Doerr, the leader of Kleiner Perkins, was the mastermind behind the Excite@Home deal when his firm owned significant stakes in both companies.) With the promotion of former Excite CEO George Bell to Jermoluk's slot, investors continued to witness a string of appalling financial reports and declines in stock price. By September 2000, Bell decided he too had had enough of the CEO job, announcing his intention to find his replacement in the CEO slot no later than the end of 2001. In the absence of some miraculous stroke of management genius, it is difficult to see how Excite@Home can become a profitable operation.

The point of these two examples is to highlight how changes in ISP corporate strategy, no matter how well conceived, can end up going in different directions depending on how well the strategies are executed. In the case of AOL, the execution to date has been good. It remains to be seen whether the Time Warner deal can pay off for investors. In the case of Excite@Home, the execution has generated such poor financial results that investors may question whether the idea of combining the two firms was good in the first place.

Brand Family

Since it is often difficult for investors to understand the details of how an ISP actually works, investors can analyze ISP's brand families as a gauge of their relative value. The conclusion of the analysis of the different firms' brand families is that there are some important differences among the firms, particularly in terms of their customers and principal investors. These differences foreshadow different capabilities of shareholder-value generation and are therefore useful indicators for investors.

The brand families of ISPs vary significantly. Investors have rewarded the ISPs with the most compelling brand families. As Table 11-5 indicates, the firms with the most prestigious customers and partners seemed to do better in the stock market than those with less prestigious customers and partners. For example, MCI WorldCom played a prominent role in Metricom's stock market success—transferring its corporate prestige in the eyes of institutional investors to Metricom through its decision to purchase preferred stock and its distribution deal to sell Metricom services. Exodus Communications attracted the most prestigious list of corporate customers and technology partners of the B2B ISPs. AOL has similarly built a powerful brand in the B2C ISP sector.

By contrast, ISPs with less distinguished brand families did less well in the stock market. For example, the partners of Excite@Home and EarthLink, while respectable, are not among the most prestigious organizations in the current Internet-oriented business environment.

TABLE 11-5. MAJOR CUSTOMERS AND PARTNERS OF SELECTED ISP COMPANIES

Company	Major Customers/Partners
Metricom	29,500 Ricochet subscribers. In November 1999, MCI WorldCom, Inc., and Vulcan Ventures each purchased shares of preferred stock for $300 million in cash. In addition, MCI WorldCom sells subscriptions to its service and has agreed to pay at least $388 million in revenue over the five years following the launch of its service.
Exodus	2,200 customers under contract and managed over 27,000 customer servers. Customers include Yahoo, USA TODAY.com, weather.com, priceline.com, British Airways, and Nordstrom. Systems vendor platforms from Compaq, Dell, Hewlett-Packard, IBM, Microsoft, Silicon Graphics, and Sun Microsystems, as well as network vendor platforms from F5 Networks, Alteon Web Systems, ArrowPoint Communications, Cisco Systems, Foundry Networks, AXENT Technologies, and Check Point Software Technologies.
Juno Online	Major customers include IBM, AT&T Wireless, Bank of America, eBay, Hewlett-Packard, Intel, Microsoft, and the U.S. Navy.
PSINet	91,000 corporate customers, 2.0 million consumers. 90 of the 100 largest metropolitan statistical areas in the U.S. have a presence in the 20 largest telecommunications markets globally and operate in 27 countries.
America Online	AOL service, with 25 million members, and the CompuServe service, with 2.7 million members; Internet brands include ICQ, AOL Instant Messenger, and Digital City; the Netscape Netcenter and AOL.COM Internet portals; Time Warner assets.
IDT	125 wholesale customers; 65,000 retail customers; Net2Phone Direct service has 1.8 million registered customers worldwide.
Excite@Home	1.1 million subscribers to @Home service and 5,100 @Work business access customers. The Excite Network has 51 million registered users, fifth-widest reach among all Internet properties according to Media Metrix; cable partners include AT&T, Comcast, Cox, Cablevision, Rogers, Shaw; content partners include Intuit, Inc., Sportsline USA, Tickets.com, and WebMD; acquisition of

Company	Major Customers/Partners
	Bluemountain.com increased reach by 50%, adding 9.6 million unduplicated users.
EarthLink	3.1 million paying members on December 31, 1999; strategic alliance with Apple, Packard Bell, NEC, and CompUSA; affinity marketing partners include Sam's Club, Discover Card, and AAA of Southern California.

Source: Company Financial Statements (10Ks)

The investor composition of the ISPs reviewed in this chapter does not provide meaningful insights for investors. As Table 11-6 indicates, ISP firms' ownership consists of a mixture of large money managers, corporate partners, and a few insiders. It may be worth noting that significant retail mutual fund managers, such as Janus, Fidelity, and Putnam, seem to invest in ISPs that perform better in the stock market and seem to avoid investing in those that do not do as well. This observation raises a question regarding the direction of causality. Simply put, it is not clear whether mutual funds' decisions to invest in these better-performing ISPs was the cause of their superior performance or whether the funds began to invest in the better-performing ISPs after that superior performance had already taken place. In 2000, the benefit of Paul Allen's investment in Metricom was overwhelmed by Metricom's dismal financial performance. The foregoing analysis suggests that potential investors should not draw significant conclusions about the upside stock price performance potential of an ISP solely on its major investors.

TABLE 11-6. MAJOR INVESTORS IN SELECTED ISPs

Company	Major Investors
Metricom	Paul Allen (49%); WorldCom (38%)
Exodus	FMR (13.4%); Janus (8.4%) K. B. Chandrasekhar (4.2%) Ellen M. Hancock (2.0%); John R. Dougery (1.5%)
Juno Online	David E. Shaw (45.3%), News America Corp. (8.1%)
PSINet	IXC Internet Services (13.4%); Janus (11.7%) William L. Schrader (7.4%), T. Rowe Price (6.6%)

Company	Major Investors
America Online	Janus (3.3%), Barclays Bank (2.8%), FMR (2.2%)
IDT	Company founder and CEO Howard Jonas owns 32%
Excite@Home	AT&T(25%) and has a 56% voting interest. Other investors include cable companies Comcast, Cox Communications, and Cablevision and Kleiner Perkins
EarthLink	Charles G. Betty (1.0%); Charles M. Brewer (3.9%); Sky D. Dayton (4.0%); Sprint (18.7%)

Source: Company Proxy Statements

Investors place a greater emphasis on an ISP's customers and partners than they do on its major investors. The relative quality of an ISP's customers and partners could be a useful indicator of the ISP's price appreciation potential. If an ISP has the best customers and partners, investors seem more inclined to buy that stock than if the ISP's customers and partners are below the top tier.

As a general rule, the ISP market seems to be one where investors should focus their attention on companies that create a virtuous cycle. By offering a service that provides the most customer value, these ISPs attract the best customers and partners. By attracting the best customers and partners, they grow the fastest and receive the capital they need to finance improvements to their service—thereby attracting more of the best customers and partners. Conversely, firms that do not offer the best customer value are unable to attract the best customers and partners, grow more slowly, and are unable to attract sufficient capital to keep up with their peers. Eventually, these second-tier firms spiral out of existence.

Financial Effectiveness

Financial effectiveness is not the most important factor that investors use to discriminate among ISPs. While sales per employee seems to be a useful benchmark—particularly if there is a distinction made between B2C and B2B—the ratio of SG&A to sales does not seem to be a useful statistic for predicting ISP stock market performance.

Sales productivity within the B2C and B2B segments does seem to be a fairly useful predictor of ISP stock market performance. As Figure 11-6 suggests, the B2C ISPs with the best sales productivity performed the best in the stock market in 1999, as did many of the B2B ISPs. More specifically, AOL, with sales per employee of $521,000, did much better in the stock market than its less productive peers, Excite@Home and EarthLink.

In the B2B sector, with the exception of Metricom and Concentric Networks, the better performing ISPs in sales productivity did better in terms of stock price appreciation. For example, Exodus generated sales of $202,000 per employee, significantly outperforming such rivals as Verio and PSINet in both sales per employee and 1999 stock price performance.

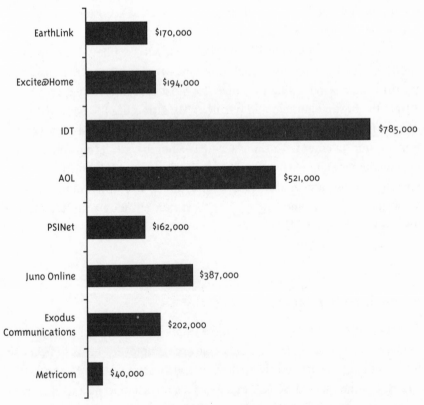

Figure 11-6. Sales per Employee of Selected ISPs, 1999
Source: MSN MoneyCentral

Similar correlations between stock price performance and relative levels of SG&A to sales do not apply. As Figure 11-7 indicates, firms with relatively high SG&A-to-sales ratios performed better in the stock market than companies with lower SG&A-to-sales ratios. The most extreme example of this is IDT, whose 18% ratio of SG&A to sales was by far the lowest among the companies analyzed here. As previously noted, IDT's stock market performance was among the weakest of the group in 1999.

This analysis suggests the possibility that investors perceive that there is an optimal level of SG&A to sales—potentially between 30% and 55%. If a firm spends less than that amount, investors may perceive the firm as

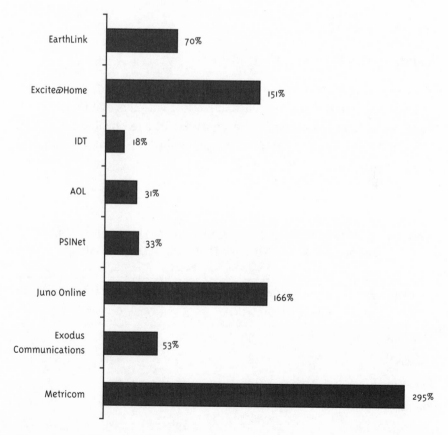

Figure 11-7. Selling, General, and Administrative Expense as a Percent of Sales for Selected ISP Firms, 1999

Source: MSN MoneyCentral

under investing in its brand. Above that amount, investors may presume that the firm is spending above the level at which it is reasonable to earn an attractive return on the investment. Simply put, it may be that for ISPs there is a level of SG&A spending that is a good investment, and any amount spent above that is perceived as a wasteful expense.

The conclusion from this analysis is that ISP investors use relative levels of sales productivity and SG&A to sales as indicators of where to invest. B2B and B2C firms with the highest sales productivity tend to do better in the stock market than those with lower sales productivity. The lessons from analysis of SG&A-to-sales ratios are somewhat murkier. It could be that investors place a higher price on firms that invest an optimal amount in SG&A to sales—punishing firms that either under- or overinvest in SG&A.

Relative Market Valuation

The stock market's approach to valuing ISP companies suggests that most of the firms appear fairly valued relative to each other with two exceptions. As Table 11-7 indicates, there appears to be a reasonable correlation between the ratio of price/sales and revenue growth for all the ISPs studied in this chapter with two exceptions:

TABLE 11-7. STOCK PRICE AND REVENUE GROWTH PERCENT CHANGE, 2000, AND VALUATION RATIOS FOR SELECTED ISPs, JANUARY 12, 2001

Company	Stock % Change	Revenue Growth	Price/Sales	(Price/Sales) /Rev. Growth	Price/Book
Metricom	−87%	−30	26.48	−88.27	1.13
Exodus	−55	287	12.72	4.43	15.99
Juno Online	−98	148	0.58	0.39	1.77
PSINet	−98	128	0.37	4.88	1.70
America Online	−54	35	16.18	46.23	16.51
IDT	8	49	0.74	1.51	0.59
Excite@Home	−87	200	5.90	2.95	0.55

Company	Stock % Change	Revenue Growth	Price/Sales	(Price/Sales) /Rev. Growth	Price/Book
EarthLink	−81	197	1.06	0.54	0.80
MEAN	**−69%**	**94%**	**8.00**	**−3.42**	**4.88**

Source: MSN MoneyCentral

Metricom appears to be overvalued, and IDT appears to be undervalued.

Metricom is trading at a price/sales ratio of 26 and had a 30% decline in revenues. Even after Metricom's significant drop in value in 2000, it still appears grossly overvalued relative to its peers. Conversely, IDT appears to be significantly undervalued. With its price/sales ratio of 0.74 and its revenues growing profitably at 49%, IDT's price/sales-to revenue ratio is a tiny fraction of the ISP sample average price/sales ratio of 15.18. Although IDT owner Howard Jonas appears to have a somewhat shaky reputation, it is not clear whether this reputation warrants such a low valuation on IDT. Perhaps there is more going on within the company than meets the eye. IDT's business certainly deserves further examination to assess whether its low valuation is warranted.

The Internet Investment Dashboard analysis suggests that ISP could be an attractive place for long-term investors who are selective. In Figure 11-8, all the factors except economic bargaining power (hatched) are dotted. Such a configuration strongly argues for a careful company-by-company analysis.

The industry is moderate in size but is growing quickly. There are many competitors and lots of consolidation. The average ISP is unprofitable because it cannot charge a price high enough to offset the costs of supporting such rapid demand growth. Industry leaders such as Exodus Communications and AOL declined less than their peers in 2000, although investors have favored B2B ISPs over B2C. Industry laggards are very unprofitable and have either announced merger partners or are likely to announce such partners.

Investors clearly place a premium on the quality of an ISP's closed-loop solution. Firms that offer really effective solutions for businesses or consumers have done better in the stock market and are likely to continue to do so. Firms that do not execute as well, particularly in the area of integrating acquisitions, tend to suffer in the stock market.

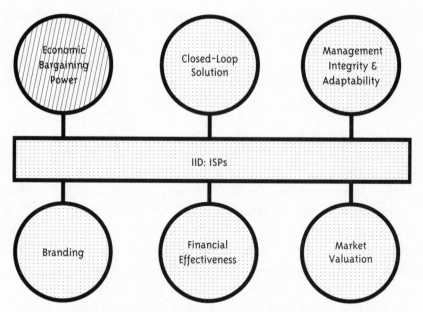

Figure 11-8. Internet Investment Dashboard Assessment of ISP Segment
Key: Dotted = mixed performance and prospects; hatched = poor performance and prospects.

Investors also place an emphasis on evaluating ISP management integrity and adaptability. Investors punish firms that demonstrate lapses in management integrity and tend to reward firms whose changes in corporate strategy lead to superior financial performance. Financial effectiveness is increasingly important to ISP investors. Investors tend to reward firms with relatively high sales productivity who spend SG&A within an optimal range. Finally, as of this writing, the market has assigned relative values to ISPs that make sense in relation to each other with the exception of Metricom and IDT. The former appears overvalued and the latter undervalued.

PRINCIPLES FOR INVESTING IN ISP STOCKS

The analysis we have conducted here is subject to change. Some of the strong-performing companies will falter, some of the weak ones could generate positive surprises, and new competitors could emerge to tilt the playing field in a totally different direction. As a result, investors in ISP stocks need some principles to guide them through the analytical process as investment conditions evolve.

ISP Investment Principle 1. Focus attention on the few ISPs with the potential to generate economic power. As we have noted earlier, the average ISP firm is unprofitable. But AOL is quite profitable because it has so many paying visitors that it has significant economic bargaining power with potential advertisers and partners. This bargaining power is likely to grow as AOL integrates Time Warner. Exodus Communications is perhaps the leading B2B ISP. Given the tremendous demand for its services and the challenges that companies face in finding Web-hosting firms that can deliver high levels of service, Exodus should ultimately be in a position to raise prices enough to offset its costs. Since Exodus is still trying to build dominant market share, it may be too soon to expect Exodus to begin raising its prices. Investors should monitor pricing trends in this segment, because if firms such as Exodus are able to raise prices, their stock prices should rise substantially.

ISP Investment Principle 2. Monitor ISPs that offer closed-loop solutions whose value is verified by customers and independent analysts. Customers of ISPs value the ability to provide a true closed-loop solution. Investors can verify a true closed-loop solution by ascertaining the views and opinions of customers and independent analysts. In the past, investors have boosted the shares of ISPs that offer such closed-loop solutions. Whether they will do so in the future is likely to depend on whether these firms can translate their effective strategies into profits.

ISP Investment Principle 3. Avoid investing in firms where there are questions about management integrity and adaptability. Our analysis indicated questions about management integrity and adaptability in a few firms. This analysis is most useful for investors as a sell signal. If a firm with a track record of avoiding shareholder lawsuits and generating consistent profit growth suddenly deviates from this path, there is a good chance that investors should consider selling the stock. In many cases, such deviations from the path of integrity and adaptability could be a signal that management is no longer up to the job. When these deviations occur, it is likely that management will try to gloss over the problems. Investors may not want to stay along for the ride.

ISP Investment Principle 4. Invest in ISPs with compelling brand families. As we noted several times in this chapter, the quality of an ISP's partners and customers sends a strong signal about the firm's relative value to potential investors and future customers. It is highly likely that an ISP that starts off in business with highly regarded customers and partners will be able to sustain this level of quality. Firms that start off with strong partners and customers and later allow the quality of their brand family to deteriorate may be sending a signal to investors that their shares should be sold. Conversely, firms whose partners and customers gain in quality over time may be signaling an increase in their share value, which could make them good investments.

ISP Investment Principle 5. Validate the effectiveness of ISP's strategies by analyzing their financial effectiveness. Investors are likely to be paying more attention to financial effectiveness among ISPs. Our analysis suggests that investors drove up the stock prices of the B2B and B2C ISPs with the highest sales productivity within their segments. Our analysis also suggested that ISPs that spend an optimal amount on SG&A are likely to enjoy the greatest stock price appreciation. While the precise levels of SG&A spending are likely to evolve over time, our analysis is likely to be useful for investors if used in conjunction with the other indicators.

ISP Investment Principle 6. Monitor relative valuations among B2B and B2C ISPs to identify potentially under- and overvalued companies. As we noted earlier, most ISPs are fairly valued relative to each other. Two firms in particular were singled out, one as possibly overvalued, the other as possibly undervalued. While these specific firms are likely to change, investors should monitor the valuation indicators we detailed earlier in this chapter. Changes in relative values among ISPs in the two sectors are likely to indicate opportunities to take advantage of valuation disparities. The stock price behavior of Metricom and IDT relative to these valuation benchmarks should provide investors with a useful guide.

CONCLUSION

ISPs may present investment opportunities for investors who analyze carefully the six IID indicators. While the average ISP lacks economic bargaining power, investors should seek out the handful that have such power. Furthermore, investors should test the quality of these firms' strategies to assess from the perspective of customers and independent analysts whether selected ISPs deliver a true closed-loop solution (as opposed to the claims of the ISP). Investors should shun firms that lack management integrity and have failed to adapt to changing customer needs, new technologies, and upstart competitors. Investors should seek out ISPs with compelling brand families that have demonstrated financial effectiveness. Finally, investors should monitor changes in relative ISP valuations to pick the best times to invest in firms that pass the other five tests.

Chapter 12

PICKS AND SHOVELS OF THE INTERNET GOLD RUSH: WEB TOOLS STOCK PERFORMANCE DRIVERS

WEB TOOLS HAVE EARNED substantial returns for investors because every e-commerce firm needs these tools to compete. As dot-coms and their land-based peers scrambled to build Web sites, they all used Web tools—regardless of their ultimate market position. This does not mean that investors in all Web tools companies have profited. The simple fact remains that there are big differences among these companies. Those that have created software that helps companies accelerate Web site development have experienced faster and more predictable revenue growth. The more rapid, and yet predictable, the financial performance, the faster the stock price appreciation. Conversely, Web tools firms that did not develop solutions that bolstered the rate at which customers could build Web sites encountered weak operating performance and declining stock prices.

BroadVision's stock market performance reflects the popularity of its products and the consistency of its financial results. BroadVision's stock appreciated 118%, from $11 to $35, from September 1999 to September 2000. BroadVision's stock suffered considerably during this period, however, having dipped 157% from its $90-per-share 52-week high reached in March 2000. As we will soon see, BroadVision has a strong list of respected customers and is expected to resume profitable operation. Nevertheless, the April 2000 Internet stock market crash has caused investors to anticipate that dot-coms and land-based firms building Web sites may not have the cash needed to develop their Web sites going forward.

The shareholders of Open Market have not fared so well. For example, Open Market stock lost 43% of its value from September 1999 to September 2000, declining from $14 to $8 per share. This decline does not capture the full extent of the damage, however, since Open Market's shares reached a 52-week high of $65 in March 2000. Investors who bought Open Market at this peak suffered an 88% drop in value. Beyond the problems related to the Internet stock crash, Open Market has experienced significant management turmoil and faced a real challenge in creating closed-loop solutions for customers. Whether Open Market can survive as an independent entity and whether it would be worth acquiring both remain to be seen.

Web tools are to the Internet era what picks and shovels were to the California gold rush of the 1850s. Web tools help organizations build, operate, evaluate, and improve their Web sites. For example, companies such as BroadVision sell software that other companies can use to build Web sites that can be customized to individual users. Vignette's software helps companies manage the content on their sites. Macromedia's software enables developers to produce vibrant digital images. Accrue's software lets companies analyze who visits what parts of their sites so they can adapt their sites to meet the interests of their target customers. Inktomi provides Web search capabilities and infrastructure services to speed the flow of content from source to destination.

With all the investment that firms have been making to participate in the Internet business, it is perhaps not so surprising that Web tools companies have experienced very rapid growth. As we will discover, the average Web tools firm is not making money. A big reason for the lack of profitability in the Web tools sector is that firms are investing heavily to sell and market their products and to set up distribution channels. Furthermore, many Web tools companies are discovering that despite the perceived need to build and improve Web sites, many organizations are still behind the curve. Thus their current and potential clients need help with the design and implementation of their Web sites before they can even begin to think about buying tools for them. As a result, Web tools companies may find that they need to form partnerships with Web consultants in order to unlock their full growth potential. Where this leaves the industry in terms of profit potential remains to be analyzed in this chapter.

WEB TOOLS STOCK PERFORMANCE

The stock price performance of the Web tools segment was outstanding—on the way up in 1999 and on the way down in 2000. We noted in Chapter 3 that the stock market performance of a cross section of Web tools firms was 448% in 1999. This average performance was over five times better than the NASDAQ's 86% increase and 25% better than the 339% average stock price increase of the companies in our nine-segment Internet stock index in 1999. But 2000's Web tools stock market performance represented a dramatic reversal of 1999's result—an 81% drop. Most of the drop is attributable to the decline of Web advertising software stocks, such as DoubleClick and Net Perceptions, which suffered as Web advertising declined.

The strong stock market performance of the Web tools sector was certainly aided by the performance of four particularly strong companies. As Figure 12-1 indicates, the four top-performing stocks were BroadVision, DoubleClick, Vignette, and RealNetworks. As we will explore later in this chapter, these firms are leaders in their respective market segments: BroadVision dominates the market for Web tools to build personalized Web sites; DoubleClick is the leader in Web advertising-management services; Vignette dominates the content-management segment; RealNetworks is the dominant provider of audio and video processing software for the Web.

The stock performance of these companies seems to hint at the relative importance of the Web tools' software to customers. For example, the software of top stock market performer BroadVision is among the most essential elements of an e-commerce architecture, whereas the stock market laggard Net Perceptions' collaborative filtering software, while useful, is not critical for the operation of many Web sites.

The April 2000 crash in Internet stocks had a profound impact on the Web tools stocks profiled in this chapter, slicing 61% off their average value. As Figure 12-2 (page 310) indicates, the crash sent a powerful message: Profits matter. Macromedia was the only profitable Web tools company, and its stock price declined a relatively modest 17%.

Other Web tools stocks declined much more due to investor concerns regarding a slowdown in dot-com demand, declining Internet advertising, and in some cases mismangement. For example, Open Market seriously disappointed investors, leading to the departure in February 2000 of its

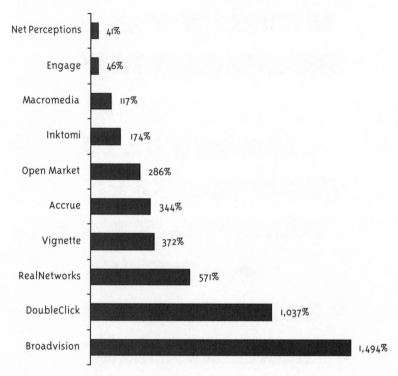

Figure 12-1. Selected Web Tools Firms' Percent Change in Stock Price,
12/31/98–12/31/99
Source: MSN MoneyCentral

CEO, Gary Eichorn. In July 2000, Eichorn's successor, Ron Matros, was also replaced. And DoubleClick found itself embroiled in serious social and political concerns about how DoubleClick was seeking to sell to advertisers information about individual Web users' surfing habits and contact information. This apparent abuse of privacy sent shudders of fear through DoubleClick investors and drove down DoubleClick's stock price.

INTERNET INVESTMENT DASHBOARD ANALYSIS OF WEB TOOLS

While the foregoing discussion of the relative stock price performance of the Web tools segment is suggestive, it does not offer a fully satisfactory explanation. To obtain such an explanation, we apply the Internet Investment Dashboard (IID) analysis to the Web tools segment.

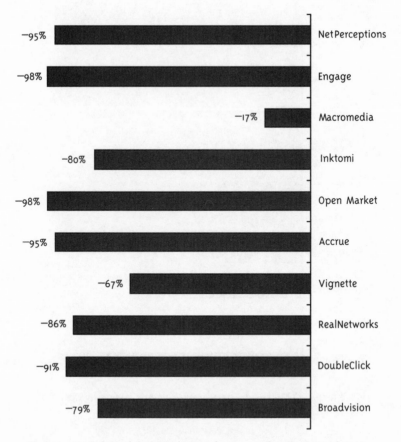

Figure 12-2. Selected Web Tools Firms' Percent Change in Stock Price,
12/31/99—12/29/00
Source: MSN MoneyCentral

As we will see, the IID analysis suggests that selected companies in specific Web tools sectors are likely to generate attractive returns for investors. As demand to build Web sites persists, the Web tools providers with leading market positions are likely to benefit from growing economic bargaining power, which will enable them to charge high prices. Web tools firms that offer customers closed-loop solutions are likely to be in the strongest position to profit from this enhanced economic bargaining power. Conversely, firms that offer point solutions may find themselves growing more slowly, ultimately leading them to merge with other firms that are seeking to assemble a full portfolio of Web tools. In fact, it is not too much of a stretch to imagine one firm that offers all the Web tools integrated seamlessly. In this chapter we will examine vendors who may be seeking to realize this goal.

ECONOMIC BARGAINING POWER

As we suggested earlier, the economic bargaining power of the Web tools industry varies depending on the strategic importance of the tools. For example, the makers of Web tools that are on the critical path for a corporate Web site—those needed for building a Web site and for linking it with a catalog of products and prices—have more economic bargaining leverage than makers of software for measuring who visits a Web site and where they visit. Simply put, a company needs to build its Web site before it can measure its performance. As a consequence, the companies that have been leaders in the market for tools to build and operate Web sites have had greater economic bargaining power than firms selling the Web site measurement and analysis tools. In between these two categories of Web tools, there are a number of other categories of tools that carry varying levels of economic bargaining power.

While these various segments enjoy differing levels of economic bargaining power, overall, the Web tools industry is large and is likely to expand rapidly, at an estimated compound annual growth rate of 96%. As Figure 12-3 indicates, it is anticipated that the worldwide Internet commerce application market will grow from $444 million in 1998 to $13 billion in 2003.

As we noted earlier, many Web tools companies are beginning to recognize that companies often need to work with Web consultants in order to conceive their e-commerce strategies and to design the architecture of their e-commerce systems. Once the e-commerce system architecture is designed, companies begin to evaluate tools to help implement the architecture. As illustrated in Figure 12-4, spending on software and services to support e-commerce alone exceeded $5.6 billion in 1998 and will grow at a 56% compound annual rate to $35 billion by 2002.

Despite the significant size of the Web tools market, the average Web tools company analyzed in this chapter was unprofitable. As Table 12-1 (page 313) illustrates, the average net margin for the 10 Web tools companies sampled in this chapter is −72%. This means that for every dollar

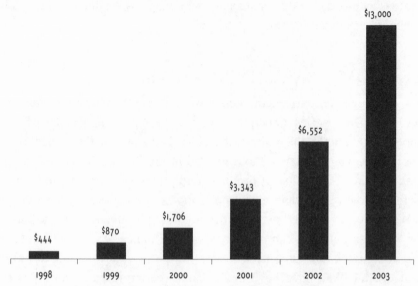

Figure 12-3. Worldwide Internet Commerce Application Market, 1998–2003,
in Millions of Dollars

Source: International Data Corporation (from WebMethods Si)

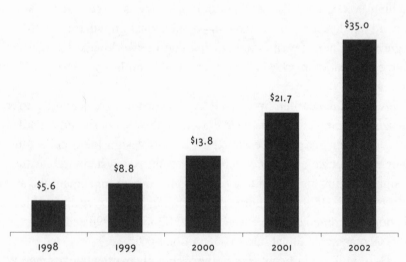

Figure 12-4. Software and Services to Support E-Commerce, 1998–2002,
in Billions of Dollars

Source: Forrester Research (from Accrue Si)

that these companies generated in revenue, they lost 72 cents. In fact, this average performance includes big money losers like Engage, Vignette, and Accrue, which lost $2.14, $0.67, and $1.05, respectively, for each dollar of revenue. On the moneymaking side was Macromedia, which earned an 8% net margin.

TABLE 12-1. MARGINS FOR 10 WEB TOOLS FIRMS, DECEMBER 2000

Company	Stock % Change	Gross Margin	Operating Margin	Net Margin	2 Year Trend
Broadvision	−79%	82	−24	−28%	−
DoubleClick	−91	69	−15	−17	+
RealNetworks	−86	76	−30	−31	−
Vignette	−67	65	−66	−66	—
Accrue	−95	84	−105	−105	—
Open Market	−98	52	−56	−56	—
Inktomi	−80	96	−3	−3	+
Macromedia	−17	97	13	8	+
Engage	−98	34	−214	−214	—
Net Perceptions	−95	70	−205	−205	−
MEAN	**−81%**	**73%**	**−71%**	**−72**	−

Source: MSN MoneyCentral

Note: Two-year trends represent the extent to which net margins have changed in the most recent two years. + indicates a small improvement, − indicates small deterioration, — indicates large deterioration.

Analyzing the reasons for the differences in profit can provide investors with useful insights. The biggest money-losers are the Web tools companies that must convince companies of the value of their products in a weak market for Web advertising. Firms like Engage and Accrue face considerable challenges in this area. While conceptually it may make sense to spend money on software to analyze which parts of a Web site are most frequently visited and how best to target the content of a Web site, these functions are not the most critical concerns facing a Web site manager. As a result, firms like Engage and Accrue, which sell products that do these things, generally take longer to make their sales and must pay their sales forces a relatively significant fixed amount of money to close these sales.

Web tools companies that make software to deal with Web site managers' most pressing concern—building a Web site that does the basics well—tend to have higher profits because their products sell themselves. More specifically, Web tools sold by Macromedia, the one profitable firm in this category, tend to be in great demand by Web site developers. Given the demand for the products, the cost of selling each incremental unit is lower than that of firms like Engage and Accrue, which must spend more time pushing their products on relatively skeptical buyers.

The Web tools industry is also subject to a powerful network effect that is likely to open up a gulf in the industry between the few profitable firms and a larger number of unprofitable ones. As we noted earlier in Chapter 2, in many Internet business segments the value of the business increases with the square of the number of product users. In Web tools, this network effect is working on overtime. For example, Macromedia makes a product called Flash, which is used by many Web sites to build visually appealing advertising. People visiting such a Web site are prompted to download Flash onto their computers in order to view the advertisement built in Flash.

Macromedia enables the individual to download the Flash reader in a fairly short time and at no cost. Once Flash is downloaded, Macromedia records the new user. Additional Flash users increase the value of programs developed in Flash by widening the market to which Flash developers will be directing their work. As a consequence, more and more developers use Flash because it is used by so many Web sites. The consequence of this virtuous cycle has been that Macromedia's market share in this category has grown rather rapidly, to almost 90%, even as the overall market has grown rapidly.

Firms that gain economic bargaining power are creating such virtuous cycles. Firms that lose economic bargaining power are unwilling victims of their competitors' virtuous cycles. Investors who can identify firms on the winning side of the power curve are likely to find profitable investments. Investors who identify firms on the losing side of the power curve may find companies that it is most useful to avoid.

Closed-Loop Solution

The Web tools sector is in a state of transition similar to what occurred in the Internet infrastructure and Web security sectors. During the early stages of industry development, vendors develop point products, which meet specific parts of an organization's business need. While these point solutions are useful, over time organizations realize that they need these point solutions to share information. Often, different vendors are selling different pieces and point solutions that do not share information effectively. As a consequence, organizations realize that a closed-loop solution, in which all the different point solutions could share information seamlessly, would be more valuable.

As organizations demands evolve from point products to closed-loop solutions, the vendors that can evolve to meet their customers' changing demands tend to generate superior financial results and to perform better in the stock market. Investors who can recognize the characteristics of these changing demands and can assess which suppliers are adapting most effectively are likely to make better investment decisions. Investors could have profited from this trend by identifying Cisco Systems as the firm that began offering closed-loop solutions in the Internet infrastructure sector in the mid-1990s. Similarly, investors who recognized the same dynamic in the Web security market could have identified Check Point Software as an emerging provider of integrated security solutions.

Now, let's explore how this pattern might apply in Web tools. One way to assess the relative effectiveness of the strategies of firms in the Web tools market is to analyze the firms' relative size, revenue growth, and market capitalization. This analysis can help us gain insights into which firms are generating the greatest product market momentum and the extent to which that momentum has translated into greater stock price appreciation.

The revenues of most of the firms analyzed in this chapter grew faster than the 96% compound annual growth rate of the Internet commerce applications market. The firms that grew more slowly than the overall average tended to perform less well in the stock market than the firms whose growth rates exceeded the average. As Table 12-2 indicates, all the Web tools stocks lost value in 2000. The firms most heavily dependent on Web advertising lost the most ground. The firms with the highest market capitalizations were more focused on helping land-based companies build e-commerce capabilities.

TABLE 12-2. REVENUE MOMENTUM OF 10 WEB TOOLS FIRMS, JANUARY 16, 2001, IN MILLIONS OF DOLLARS

Company	Stock % Change	12-Months Revenue	Revenue Growth	Market Capitalization
Broadvision	−79%	$321	286%	$4,260
DoubleClick	−91	534	281	1,820
RealNetworks	−86	227	109	1,620
Vignette	−67	284	404	3,170
Accrue	−95	32	196	87
Open Market	−98	104	41	131
Inktomi	−80	171	222	2,110
Macromedia	−17	350	83	2,220
Engage	−98	177	1,005	258
Net Perceptions	−95	36	236	67

Source: MSN MoneyCentral

Let's look at some examples of these general trends. With a couple of exceptions, investors clearly put the highest market value on the fastest-growing Web tools firms in 1999. By January 2001, BroadVision had grown revenues 286%, and its market capitalization had grown to over $4.2 billion. DoubleClick was actually larger than BroadVision in terms of sales and grew at almost 281% in 2000; however, in the first half of 2000 DoubleClick stock tumbled 200% from its 1999 high, ending with less than half the market capitalization of BroadVision, at $1.8 billion. As we noted in analyzing Table 12-1, investors seemed to factor relative levels of economic bargaining leverage into their assessment of the relative value of the stocks in the sector. For example, DoubleClick's stock has suffered from investors' concerns about regulatory efforts to limit DoubleClick's ability to generate revenues from the sale of data about Web surfers' behavior—a powerful factor in reducing DoubleClick's economic bargaining power with advertisers.

This assessment of relative economic bargaining leverage may help explain why firms such as Engage and Net Perceptions fared relatively poorly in the stock market despite their higher revenue growth rates. For example, Engage, with its 1,005% revenue growth, saw its 1999 stock price appreciate a relatively low 94% in 1999. In 2000, the stock declined 98% from its 1999 year-end level, ending the period with a market capitalization of $258 million following an announcement to cut half its staff in 2001. Net Perceptions increased its revenues 236% and its stock price rose 41% in 1999, then declined 95% in 2000, and ended the period with an even smaller market capitalization of $671 million. Simply put, these firms had higher revenue growth rates but did much worse in the stock market than their peers. The relatively weak stock market performance is tied to the huge net losses both firms suffered, which were driven by the high cost of selling their products and declining growth in Web advertising.

In order to gain meaningful insights that can explain how the differences in strategies of Web tools firms translate into different levels of return in the stock market, it is important to analyze the different strategies of the firms profiled here. As Table 12-3 indicates, the Web tools firms with the greatest relative strengths and least severe competitive weaknesses tended to perform better in the stock market. Conversely, firms with greater competitive vulnerabilities and fewer relative strengths tended to perform less well.

TABLE 12-3. STRATEGIC DIFFERENCES AMONG 10 WEB TOOLS FIRMS

Company	Competitive Strengths/Weaknesses
Broadvision	*Strengths:* One-To-One products enable users to manage online transactions, order fulfillment, billing, and customer service; also lets users collect, track, and manage information about Web site visitors and use profiles to customize content. *Weaknesses:* Competitors gaining significant market strength; unprofitable.
DoubleClick	*Strengths:* Leader in ad serving with 1,500 Web sites using DART technology (DART collects and analyzes audience behavior and predicts which ads will be most effective, measures ad effectiveness, and provides data for Web publishers and advertisers). *Weaknesses:* Dependence on AltaVista (about 10% of sales), Travelocity, and Egghead.com—all in relatively weak financial condition; government concerns about privacy could affect revenue.

Company	Competitive Strengths/Weaknesses
RealNetworks	*Strengths:* RealPlayer and RealJukebox software let users locate and listen to audio and view video over the Internet; 80 million downloaded software; strong management team. *Weaknesses:* Unprofitable.
Vignette	*Strengths:* Vignette's StoryServer enables businesses to post information on their Web sites, generate reports on Web site traffic, and personalize Web pages. *Weaknesses:* Operating losses; competition.
Accrue	*Strengths*: Accrue Insight analyzes Web traffic characteristics such as visit frequency and duration. *Weaknesses:* Unprofitable, drop in stock price, open-loop solution.
Open Market	*Strengths:* Software manages transactions; electronic catalogs; Web content. *Weaknesses:* Poor financial performance, lost market share, management turnover, poor stock market performance.
Inktomi	*Strengths:* Traffic Server increases network speeds and is licensed to ISPs; Content Delivery Suite reduces content delivery time; search engine software; wireless alliances. *Weaknesses:* Unprofitable; competition in all product segments.
Macromedia	*Strengths:* Online publishing tools are market leaders; deconsolidated unprofitable shockwave.com; strong management. *Weaknesses*: Profitable, but margins dropped.
Engage	*Strengths:* Software lets clients collect profiles of Internet users to target online advertising. *Weaknesses:* Weak financial performance; regulatory threat to business; excessive dependence on CMGI for capital; competition.
Net Perceptions	*Strengths:* Collaborative filtering software offers shopping suggestions. *Weaknesses:* Poor financial performance; some clients in weak financial condition.

Source: Company Reports, Peter S. Cohan & Associates Analysis

Let's examine how these principles play out in practice. The Web tools companies with the best stock market performance tended to offer more components of a closed-loop solution than the weaker-performing firms. For example, BroadVision's software incorporates far more links in the e-commerce value chain than Net Perceptions' software. More specifically, BroadVision lets an e-commerce provider manage online transactions, fulfill orders, conduct billing, collect and manage information about Web site use, and analyze that information to customize the Web site content to each Web site visitor. By contrast, Net Perceptions' software enables e-commerce providers to do just one of those activities—customizing content to each individual user based on prior purchase behavior of individuals with similar preferences. Simply put, the more links in the value chain that a Web tools company provides to an e-commerce provider, the more the stock market seems to value that Web tools company's stock.

The implications of this analysis for investors are clear. The stock market seems to reward firms that have been most successful in making the transition from open- to closed-loop e-commerce application solutions. Investors are likely to profit most from identifying specific firms that are leading this transition. BroadVision is a case in point, insofar as it offers products to customers that incorporate more links in the value chain than many of its peers. To the extent that other firms in the Web tools industry begin to broaden their product offerings to incorporate more of these links in the value chain, investors should focus on them. If these firms in transition begin to gain a significant following among customers and industry analysts, they may represent an excellent investment opportunity.

Management Integrity and Adaptability

Management integrity and adaptability are important to investors in Web tools companies. An analysis of the revenue accounting policies and lawsuits for Web tools firms in Table 12-4 indicates that all firms in the sector follow the proper procedures for revenue accounting; however, some firms have been involved in legal proceedings that raise questions about management integrity.

The firm that appeared to have the greatest legal troubles seemed to have been hurt the most by investors. As noted earlier, DoubleClick announced in February 2000 that the FTC was investigating the company for possible misuse of information on Internet users. Investors recognize that this investigation could cut off a significant source of DoubleClick's revenue and therefore represents a considerable threat to its business model. As noted in Figure 12-2, this threat has sliced a big chunk out of DoubleClick's market capitalization. While other Web tools firms have been involved in legal proceedings, these proceedings do not appear to have had as significant an impact on the other firms as on DoubleClick.

Open Market's legal problems also raise questions about its management's integrity. In fact, it is possible that there is a correlation between these lawsuits and the February 2000 departure of former CEO Gary Eichorn. The June 2000 complaints allege that Open Market made misleading statements regarding its products and competitive position beginning in November 1999. Open Market's stock value has taken a beating as illustrated in Figure 12-2—dropping 98% in 2000.

TABLE 12-4. REVENUE ACCOUNTING POLICY AND LAWSUITS FOR WEB TOOLS FIRMS

Company	Revenue Accounting Policy	Legal Proceedings
Broadvision	Recognizes software license revenues when a noncancellable license agreement has been signed and the customer acknowledges an unconditional obligation to pay, the software product has been delivered, there are no uncertainties surrounding product acceptance, the fees are fixed, and collection is probable.	Patent infringement case filed against Art Technology Group was settled by granting ATG a license in exchange for $8 million from ATG.
DoubleClick	Revenues are recognized in the period the advertising impressions are delivered, provided collection of the resulting receivable is reasonably assured.	Complaint that NetGravity directors breached their fiduciary duties in connection with the merger with DoubleClick. FTC is conducting an inquiry into its business practices to determine

Company	Revenue Accounting Policy	Legal Proceedings
		whether, in maintaining information on Internet users, DoubleClick engaged in unfair or deceptive practices.
RealNetworks	Revenue from software license fees is recognized upon delivery, net of an allowance for estimated returns.	Patent-infringement complaint; Left Bank seeks 30% of RealNetworks' revenues from the use of RealAudio technology.
Vignette	Same as BroadVision.	None
Accrue	Recognize license revenue, net of estimated returns allowance, upon product shipment.	None
Open Market	Same as BroadVision. If an acceptance period is required, revenues are recognized on customer acceptance.	Complaints allege false statements regarding current and future products and competitive position.
Inktomi	Per-query, hosting, and maintenance fees revenues are recognized in the period earned, and advertising revenues are recognized in the period that the advertisement is displayed.	None
Macromedia	Recognizes revenue from product sales at shipment based on the fair value of the element, provided that collection of the resulting receivable is probable.	Complaint alleging false or misleading facts regarding financial results and prospects.
Engage	Revenues from software product licenses, Knowledge database services, and Web site traffic audit reports are recognized in the same way as BroadVision.	None
Net Perceptions	Same as BroadVision.	None

Source: Company 10Ks

In light of the rapid rate at which the competitive environment is changing, the ability of managers to adapt to change effectively is a crucial factor in determining the long-term value of a Web tools company. To analyze this factor, we can examine the variance of percentage changes in quarterly earnings. Simply put, if a Web tools firm is able to increase profits fairly consistently in each quarter, its management is good at adapting to change. Conversely, if a firm's quarterly performance jumps around significantly, its management team is probably less effective at adapting to change. The important point here is that the ability to adapt effectively to change in a rapidly changing environment can be measured by management's ability to produce fairly consistent financial results.

In the case of Web tools firms, it is clear that this measure is somewhat useful. As Figure 12-5 indicates, consistent earnings growth is a relevant measure for only two of the 10 firms profiled in this chapter because only two of the companies actually generated positive earnings in the five quarters analyzed. The firm with the lowest earnings growth variance, Broad-Vision (12%), happened to enjoy the best stock market performance in 1999—a performance that it was unable to sustain in 2000, losing 79% of its value. Macromedia, which had the second-lowest earnings change variance (23%), did less well relative to its peers in 1999; however, in 2000, Macromedia's stock price lost a relatively small 17%, the best performance in its peer group.

Other Web tools firms either lost money during the five quarters or fluctuated between negative and positive earnings per share. As a result, the measure of earnings variance is relevant because it suggests to investors specific stocks to avoid.

Management integrity and adaptability are most helpful for investors seeking specific Web tools firms on the margins. More specifically, if a Web Tools firm has a significant legal problem, as with DoubleClick, investors should avoid the stock. If a Web tools firm has consistent earnings growth, as in the case of Macromedia, investors should consider buying the shares.

Brand Family

Since it is often difficult for investors to understand the details of Web tools process, investors can analyze the firms' brand families as a gauge of

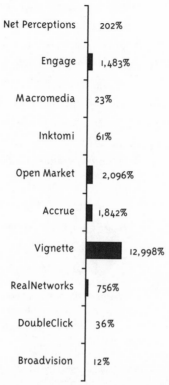

Figure 12-5. Variance in Quarterly Earnings' Percent Change for Selected
Web Tools Firms, 3/99–3/00
Source: MSN MoneyCentral, Peter S. Cohan & Associates Analysis

the firms' relative value. The conclusion of the analysis of the different firms' brand families is that there are some important differences among the firms, particularly in terms of their customers and principal investors. These differences foreshadow different capabilities of shareholder-value generation and are therefore useful indicators for investors.

The brand families of Web tools companies vary depending on whether they partner or sell directly to corporations or whether they sell their products primarily through third parties. As Table 12-5 indicates, some of the firms that form partnerships or customer relationships directly seem to have stronger brand families than firms that deal through distributors. For example, BroadVision has formed customer relationships with a significant number of global financial institutions and large retailers. Vignette has formed 515 customer relationships, including significant companies in a range of industries. Accrue has formed customer relationships with many of the leading technology companies in the world.

By contrast, other Web tools companies have relationships with decidedly second-tier customers and partners. For example, Open Market, which started in the Web tools business long before its competitors, has far fewer customers and has seen the quality and number of its customers erode. Net Perceptions, while gaining a far larger number of customers (170), depends on many second-tier firms, including dot-com companies such as Furniture.com and CDNOW, which are likely to have more trouble paying their bills.

TABLE 12-5. MAJOR CUSTOMERS AND PARTNERS OF SELECTED WEB TOOLS COMPANIES

Company	Major Customers/Partners
Broadvision	BroadVision software is used by 50 financial institutions in 25 countries. Other customers include Wal-Mart Stores, Sears, Home Depot, Federated Stores, Circuit City, and Office Max.
DoubleClick	DoubleClick is used by thousands of advertisers, advertising agencies, Web publishers, and e-commerce merchants, including AltaVista, the Dilbert Zone, Kelley Blue Book, and Macromedia.
RealNetworks	Real Networks has 700 members in its RealPartner Program. Microsoft accounted for approximately 8% of 1999 revenues.
Vignette	Vignette won the Crossroads A-list Award and *Red Herring* magazine's Top 50 Public Companies and Best Product Award. It has 515 clients, including American Express, AT&T, Bank One, Charles Schwab and Company, *Chicago Tribune,* Lands' End, Merrill Lynch, National Semiconductor, Nokia, Preview Travel, Siebel Systems, Snap.com, StarMedia Network, Sun Microsystems, and TheStreet.com.
Accrue	Clients include AMP Incorporated, Ameritrade, Apple Computer, NationsBank Corporation, DreamWorks, Check Point Software, T. Rowe Price Associates, and Eastman Kodak, Ford, Merck, Motorola, Oracle, Silicon Graphics, Sun Microsystems, and VeriSign.
Open Market	80 commerce service providers including *Business Week, USA Today,* NTT, AT&T, and First Union.

Company	Major Customers/Partners
Inktomi	Pursues opportunities with large accounts through a direct sales force, and penetrates targeted market segments through indirect distribution.
Macromedia	Macromedia's player technologies, Shockwave and Flash, are used by75% of all people on the Web; Shockwave and Flash are also distributed with Windows 95, Windows 98, Microsoft Internet Explorer, and the Apple OS. In addition, Flash is included with every AOL client, in all Netscape browsers, in WebTV, in RealNetworks' Real Player, and it has recently been licensed by @Home.
Engage	Engage had 335 customers as of July 1999. Sales to Informix Corporation accounted for approximately 12% of Engage's total revenues in fiscal 1999.
Net Perceptions	Partnerships with Vignette, IBM, Broadvision, Commercialware, Trans Cosmos, NTT Software, and Toyo Information Systems. Joint marketing relationships with Oracle, net.Genesis, and Verbind. 170 customers, including BarnesandNoble.com, Bertelsmann, CDNOW, Virgin Online Megastore, Egghead.com, Fingerhut, J. C. Penney, Dean & Deluca, Kraft Foods, Furniture.com, Procter & Gamble, and Publisher's Clearing House.

Source: Company Financial Statements (10Ks)

The major investors in Web tools companies do not appear to play an important role in influencing their stock price performance. As indicated in Table 12-6, many Web tools companies are owned heavily by their founders, often balanced with large mutual funds. This ownership structure may in fact be typical of software companies, which tend to be somewhat less capital-intensive than other Internet sectors such as Internet infrastructure. As a result of this lower level of capital intensity, many Web tools companies were heavily financed by their founders, who have thus been able to retain a larger equity share of their companies. In many other Internet businesses, venture capital firms tend to be among the largest equity holders.

In fact, investors seem to have been less interested in the Web tools firms that did take significant amounts of venture capital. Certainly, firms like Open Market, Engage, and Net Perceptions, all of which have performed less well in the stock market, took significant capital from venture firms such as Greylock Ventures, CMGI, and Hummer Windblad, respectively. Whether the dependence on venture capital and weaker relative stock price performance is a coincidence or a causative factor is difficult to determine. Nevertheless, it appears that Web tools firms in which the founders retain a significant equity stake did better in 1999 than firms that depended more heavily on outside capital. As noted earlier, investors focused more on profitability and dot-com dependence to decide which Web tools stocks to sell.

TABLE 12-6. MAJOR INVESTORS IN SELECTED WEB TOOLS COMPANIES

Company	Major Investors
Broadvision	Pehong Chen (22.2%)
DoubleClick	Kevin J. O'Connor (8.1%), Janus (10.2%), FMR (5.6%)
RealNetworks	Robert Glaser (34.6%), Mitchell Kapor (4.3%), James W. Breyer (3.2%), Phillip Barrett (2.7%), Bruce Jacobsen (1.3%)
Vignette	Putnam (5.84%), FMR (13.58%), Gregory A. Peters (3.18%), Michael J. Vollman (1.19%)
Accrue	Davidow Ventures (13.7%), Organic Holdings (9.3%), Sterling Payot Capital (9.6%), Vertex Technology Fund (6.5%), Richard D. Kreysar (6.0%), Bob Page (2.6%), Simon Roy (1.1%)
Open Market	Gulrez Arshad (1.0%), Shikhar Ghosh (7.7%), William Kaiser (4.8%), Gary Eichhorn (3.0%)
Inktomi	Eric A. Brewer (7.3%), Paul Gauthier (6.1%), Fredric Harman (3.9%), David C. Peterschmidt (3.2%), Richard B. Pierce (1.4%)
Macromedia	Putnam Investments (10.5%), James R. Von Ehr, II (7.1%), Capital Guardian Trust Company (5.0%), Geocapital (5.1%), Robert Burgess (1.6%)
Engage	CMGI (37%), Daniel Jaye (1.16%)
Net Perceptions	London Pacific Life & Annuity Company (9.9%), Norwest Bank Minnesota (9.5%), Steven J. Snyder (6.5%), Bradley N. Miller (4.8%), John T. Riedl (3.4%), P. Stephen Larsen (1.5%), Ann L. Winblad (1.2%)

Source: Company Proxy Statements

It may be that investors recognize that Web tools companies that can reach sufficient scale to go public without venture capital are simply better managed. Therefore, if the founders who were able to build their companies up retain a significant ownership share, it could imply that those firms are likely to continue to outperform their peers who were more dependent on management hired from outside.

Investors can use branding as a useful indicator in deciding which Web tools companies to buy and which to sell or avoid. The foregoing analysis suggests that Web tools firms with the best customers and partners and that retain their founders as significant shareholders tend to have better prospects. Conversely, Web tools firms that are highly dependent on outside capital and whose customers and partners are not of the first tier tend to perform less well in the stock market.

Financial Effectiveness

Investors use financial effectiveness as an important factor in deciding which Web tools firms' stocks to purchase. Rather than using sales per employee as the key financial-effectiveness measure, Web tools investors seem to use the SG&A-to-sales ratio to make choices among the companies. Figure 12-6 helps illustrate the idea that investors do not place a great emphasis on sales per employee when deciding which Web tools stocks to buy. For example, on a sales-per-employee basis, BroadVision, the top-performing Web tools stock, was only eighth best. Conversely, Open Market, a very weak stock market performer, had by far the best sales per employee at $693,000.

Despite these results, the second-best sales-per-employee performer, Macromedia ($478,000 per employee), was a top Web tools stock market performer if first half 2000 results are taken into account. Another interesting result is that Net Perceptions, by far the worst-performing Web tools firm in terms of stock market valuation, also had the worst sales per employee ($86,000).

The key to understanding how investors in Web tools companies measure financial effectiveness is to analyze their relative SG&A-to-sales performance. As Figure 12-7 (page 329) indicates, the Web tools firms with the highest SG&A-to-sales ratios tend to perform worse in the stock

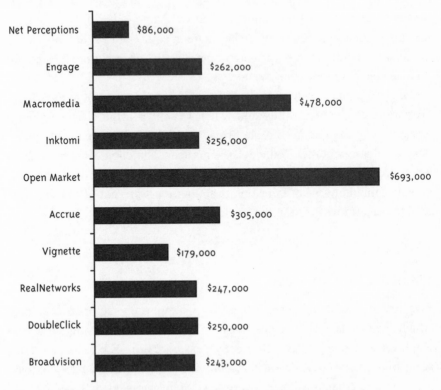

Figure 12-6. Sales per Employee of Selected Web Tools Firms, 1999
Source: MSN MoneyCentral

market than firms with lower SG&A-to-sales ratios. For example, the top three Web tools stock market performers had SG&A-to-sales ratios below 65%, while two of the worst stock market performers had SG&A-to-sales ratios of 169%.

As we noted earlier, these SG&A-to-sales ratios may be a good indicator of where various categories of Web tools fall on the spectrum of economic bargaining power. Web tools firms whose products are bought rather than sold (e.g. BroadVision, DoubleClick, and Vignette) tend to have lower SG&A-to-sales ratios. Conversely, Web tools firms whose products are sold rather than bought (e.g., Accrue, Engage, and Net Perceptions) tend to have higher SG&A-to-sales ratios.

Financial effectiveness is likely to be of increasing importance as an indicator of future investment value in the Web tools sector. The reason for

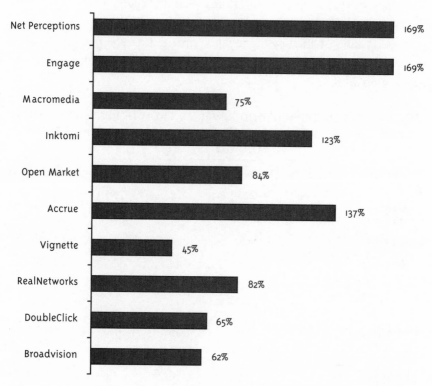

Figure 12-7. Selling, General, and Administrative Expense as a Percent of Sales for Selected Web Tools Firms, 1999
Source: MSN MoneyCentral

the emphasis on financial effectiveness is that this measure is likely to be of increasing importance to customers as the Web tools industry makes the transition from open- to closed-loop solutions. Customers will be more and more reluctant to risk the success of their e-commerce operations on a company that lacks the ability to manage its financial affairs effectively. As a result, Web tools firms that do a relatively poor job of managing their financial affairs will lose significant customers to Web tools firms that are relatively strong at financial management. The implications for investors should be clear: Consider investing in Web tools firms with improving financial effectiveness, and steer clear of Web tools firms whose financial effectiveness is eroding.

Relative Market Valuation

The stock market's approach to valuing Web tools companies reveals wide disparities among the firms. As Table 12-7 indicates, on the basis of price/sales ratio divided by the revenue growth rate, the market is assigning valuations that suggest some potentially over- and undervalued Web tools companies.

TABLE 12-7. STOCK PRICE AND REVENUE GROWTH PERCENT CHANGE, 2000, AND VALUATION RATIOS FOR SELECTED WEB TOOLS FIRMS, JANUARY 16, 2001

Company	Stock % Change	Revenue Growth	Price/Sales	(Price/Sales) /Rev. Growth	Price/Book
Broadvision	−79%	286%	13.43	4.70	3.99
DoubleClick	−91	281	3.54	1.26	2.19
RealNetworks	−86	109	7.68	7.05	3.42
Vignette	−67	404	8.94	2.21	1.48
Accrue	−95	196	2.28	1.16	0.44
Open Market	−98	41	1.36	3.32	2.88
Inktomi	−80	203	8.66	4.27	2.44
Macromedia	−17	83	6.36	7.66	7.97
Engage	−98	1,005	1.10	0.11	0.23
Net Perceptions	−95	36	1.46	4.06	0.31
MEAN	**−81%**	**264%**	**5.48**	**3.58**	**2.54**

Source: MSN MoneyCentral

More specifically, on this basis, Macromedia and Real Networks could be significantly overvalued. For example, Macromedia's price/sales ratio divided by revenue growth rate is 7.66, over twice the Web tools sample mean of 3.5. RealNetworks' price/sales ratio divided by revenue growth rate is the second loftiest at 7.05. Macromedia's relatively high valuation can perhaps be justified by its profitability. RealNetwork's high valuation may be more difficult to understand given its −31% net margin.

While there are several firms profiled in this chapter whose price/sales ratio divided by revenue growth rate is lower than the sample mean, it is not obvious that these lower-value firms are undervalued. Simply put, many of the lower-valued firms probably deserve their lower valuations. More specifically, DoubleClick's price/sales ratio divided by revenue growth rate of 1.26 is much lower than the Web tools average. Given its shaky position with regulators, however, it may be that DoubleClick is grossly overvalued even at its relatively low price. However, DoubleClick is also suffering from an anticipated slowing in Web advertising growth—making it difficult to revive its business. If it does, then DoubleClick would probably be a screaming buy at current valuation levels. As of this writing, it is difficult to see how this optimistic outcome could take place.

Engage and Accrue are also both valued at a much lower level than their peers. However, the question again arises as to whether these valuations are appropriate. If these firms can tighten their financial management and broaden their product lines to offer customers more closed-loop solutions, then these two firms may well be undervalued at current levels. It is also possible, however, that these firms could be in the middle of a vicious cycle.

This vicious cycle could operate as follows. Engage and Accrue have many dot-com customers that are likely to experience significant financial problems. The financially weaker dot-com customers may not be able to purchase additional products from Engage and Accrue and may even have difficulty paying for products that they have already purchased. As a result, Engage and Accrue may find their financial effectiveness and profit margins deteriorating even further. This could encourage larger corporate customers to shy away from doing more business with Engage and Accrue as these customers worry about whether these suppliers will have the financial stability to adapt to changing customer needs.

As a result of this potential challenge in getting current corporate customers to buy new products and the difficulty of getting new corporate customers, Engage and Accrue could find themselves losing even more revenue. This could precipitate a downward spiral, the ultimate effect of which could be further declines in stock price and ultimately liquidation or acquisition by a more financially stable company.

The Internet Investment Dashboard analysis suggests that Web tools could be an attractive area for selective investors. In Figure 12-8, all six of the factors are dotted. While most Web tools companies are losing money, one enjoys significant economic bargaining power and is profitable. Some

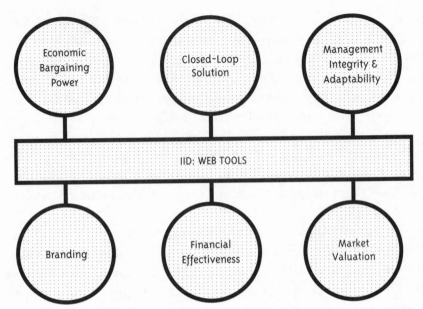

Figure 12-8. Internet Investment Dashboard Assessment of Web Tools Segment
Key: Dotted = mixed performance and prospects.

Web tools companies offer closed-loop solutions; others offer open-loop solutions. The providers of closed-loop solutions are likely to win over the long term. Most Web tools companies have management integrity, and a few adapt effectively to change.

The brand families of some of the Web tools companies are strong, but many are of the second tier. Many Web tools companies are not financially effective, and this lack of financial effectiveness is likely to harm their long-term viability. Finally, there are a few Web tools firms that appear overvalued relative to their future cash-generation potential—this, despite their strong positions in the product markets.

PRINCIPLES FOR INVESTING IN WEB TOOLS STOCKS

The analysis we have conducted here is subject to change. Some of the strong-performing companies will falter, some of the weak ones could generate positive surprises, and new competitors could emerge to tilt the playing field in a totally different direction. As a result, investors in Web tools stocks need some principles to guide them through the analytical process as investment conditions evolve.

Web Tools Investment Principle 1. Assess which firms are leading the transition from open- to closed-loop Web tools solutions. As we noted throughout this chapter, Web tools firms that provide their customers with the full set of e-commerce value chain linkages tend to perform better in the stock market than firms that offer only one or two of these links. As customers make the transition from purchasing point solutions to purchasing closed-loop solutions, firms that lead this transition will have greater economic bargaining power in their dealings with customers. Firms with such power will be in a position to negotiate higher prices and will thus make better investments.

Web Tools Investment Principle 2. Identify the Web tools firms that are able to implement closed-loop solutions that generate satisfied customers. While most Web tools firms will articulate the strategic intent to offer closed-loop solutions, what matters most for investors is whether customers find these solutions effective. Investors will profit from researching customer satisfaction and analysts' evaluations (by reading such evaluations in the trade press) to identify which Web tools firms are offering closed-loop solutions that customers are finding genuinely valuable. Conversely, investors will also find it helpful to identify firms where there is a significant gap between their stated strategic intentions and their performance in the eyes of customers and analysts. Investors should consider purchasing the stocks of Web tools firms that succeed in satisfying customers and impressing analysts and should shun the stocks of Web tools firms that fall short in this regard.

Web Tools Investment Principle 3. Avoid investing in firms where there are questions about management integrity and adaptability. Our analysis indicated questions about management adaptability in many firms. Simply put, Web tools firms with inconsistent financial performance will find it increasingly difficult to go to the capital markets to seek solutions to their cash shortfalls. Conversely, firms that can generate profits consistently will be in a stronger position to raise capital from investors. This relative ease at raising capital will increase the number of investors in those companies that can adapt effectively to change. The net effect will be a virtuous cycle, in which the rising stock prices of the few Web tools firms that adapt effectively to change will enable these winning firms to acquire the weaker firms. Investors should also be wary of firms such as DoubleClick, under investigation, insofar as such investigations create a halo of concern around the viability of the companies.

Web Tools Investment Principle 4. Assess the extent to which Web tools firms are able to create virtuous cycles in their branding. As we noted several times in this chapter, investors recognize that Web tools firms that can attract top customers and partners are in the strongest position to attract even more great customers and partners. Conversely, firms with second-tier customers and partners tend to fall by the wayside as they lose market momentum and face an increasingly difficult struggle to attract new customers. Investors should consider buying stocks in firms that are on the right side of this power curve and shun stocks in companies that are falling further and further behind. In addition, as we noted earlier, investors should consider buying stocks in Web tools companies whose owner/managers enjoy significant equity stakes. Of course, this recommendation depends on whether these owner/managers can continue to provide market leadership even as their companies reach the size where their management skills may be challenged.

Web Tools Investment Principle 5. Use financial-effectiveness measures to distinguish between long-term survivors and liquidation/acquisition candidates. Investors seem to use SG&A-to-sales ratios as a way to distinguish between the long-term survivors in the Web tools market and the firms that could ultimately be candidates for liquidation or acquisition. The firms with low SG&A-to-sales ratios seem to have greater economic bargaining power because they are selling products that customers really want to buy. The firms with high SG&A-to-sales ratios seem to be in a position where they need to hire more expensive sales forces because customers need to be convinced to buy their products. This convincing process is expensive and makes it more difficult for companies in the latter group to earn a profit. Ultimately, investors are likely to be better off investing in the firms with lower SG&A-to-sales ratios, insofar as these firms are likely to gain the lion's share of the growth in the Web tools market as their weaker competitors fall by the wayside.

Web Tools Investment Principle 6. Monitor relative valuations among Web tools sectors and within Web tools sectors to identify potentially under- and overvalued companies. As we noted earlier, there are three firms whose stock valuations appear high relative to their peers in terms of their ratios of price/sales to revenue growth rate. Firms such as BroadVision, Inktomi, and RealNetworks are all leaders in their respective markets. They all face competitive challenges, and these challenges could erode their relative market leadership and diminish the present value of their future cash flows. Investors should monitor these firms to assess whether competitors are cutting their value, and hence whether their stock prices are too high.

CONCLUSION

Web tools is an area where discerning investors can earn attractive returns. In 1999, the average stock price in the Web tools segment increased 448%. However, 81% of that value was knocked out of the group in the first half of 2000. This value-decimation process helped investors sort out the firms (such as Macromedia) that are likely to survive into the next round and those that could perish (such as Open Market). Investors who identify the Web tools firms likely to lead the transition from open- to closed-loop solutions have the best chance of profiting in the future.

Chapter 13

INVESTING IN INTERNET STOCKS

IF YOU HAVE MADE IT THIS FAR, you are probably wondering what to do with all this analysis. More specifically, you may be asking yourself the following questions:

- Should I invest in Internet stocks?
- If so, should I invest in an Internet stock index or a subset of the nine Internet business segments?
- How can I decide which specific Internet companies in which to invest?
- How can I determine the best timing to buy or sell Internet stocks?
- How can I monitor my portfolio of Internet stocks over time to determine whether to sell or buy?

We'll conclude this book by addressing these questions.

SHOULD YOU INVEST IN INTERNET STOCKS?

The answer to this question depends on who you are. More specifically, it depends on your current financial situation, your vision of where you would like to be financially, and how much risk you are willing to take to achieve your financial goals. Furthermore, it is worth emphasizing that past performance of any investment is no predictor of its future performance. Simply put, an investment strategy is formulated based on assumptions about the future which could easily be wrong.

To deal with this uncertainty, it is wise to examine different scenarios for each investment strategy. As we evaluate the questions raised at the beginning of this chapter, we will use three different forecasts—a base case, an optimistic case, and a pessimistic case. While we cannot be certain which of these cases is the most likely—or even whether the actual outcome will come close to any of these scenarios—we can be more confident that an evaluation of various investment strategies under three different scenarios will force us to stretch our thinking about the potential risks and rewards of investing in Internet stocks.

A final caveat: I believe the ideas that follow can be helpful in thinking about a financial plan; however, I am not a certified financial planner, and you should not construe what follows as a financial plan. If you want financial planning advice, seek out a highly recommended certified financial planner.

To simplify the answer to the question of whether you should invest in Internet stocks, we will discuss three types of investors: middle income, upper middle income, and affluent. We will assume that the wage earners in each family are 40 years old. The assumptions regarding each investor type's current financial condition, future goals, and risk preferences of each group are summarized in Table 13-1. Clearly, these scenarios are oversimplified. For example, the scenario could be made more robust by adding different age groups and different risk tolerances within each investor type. These scenarios also exclude costs such as taxes and transaction fees, which are likely to vary by investor.

TABLE 13-1. THREE INVESTOR TYPES

	Middle Income	Upper-Middle Income	Affluent
Current Net Worth	$100,000	$200,000	$1,500,000
Current Annual Income	$45,000	$100,000	$400,000
Future Net Worth Goal	$1,000,000	$3,000,000	$10,000,000
Future Income Goal	$30,000	$60,000	$250,000
Risk Tolerance	Willing to risk 20% of net worth	Willing to risk 30% of net worth	Willing to risk 40% of net worth

One of the key differences between the investor types is that those in the affluent group have access to a wider range of investor options than those in the other two classes. More specifically, affluent investors (as defined here) are able to invest in private equity investments, including pre-IPO Internet companies. While such investments had very attractive returns prior to the April 2000 Internet stock market crash (as illustrated in the Noosh example in Chapter 6), their potential returns are now more difficult to assess. The middle-income and upper-middle-income investors can invest in Internet index funds, Internet mutual funds, or individual Internet stocks. If they have the appropriate skills, they could also work for the Internet companies themselves, thereby getting stock options.

For the purpose of this analysis, let's assume that investors in general can put their money into a combination of three investments: a money market fund, an S&P 500 index, and an Internet stock index. Let us further assume that over a 25-year period, as a base-case scenario, the money market fund will earn a 1% annual real (e.g., inflation-adjusted) return, the S&P 500 index will earn a 7% annual real return, and the Internet Index fund will earn a 22% annual real return. Table 13-2 summarizes our assumptions regarding how these returns might vary under optimistic and pessimistic scenarios.

TABLE 13-2. ASSUMED RATES OF RETURN FOR THREE INVESTMENT CLASSES UNDER BASE-CASE, PESSIMISTIC, AND OPTIMISTIC SCENARIOS

	Base Case	Pessimistic	Optimistic
Money Market Fund	1%	0%	1%
S&P 500 Index Fund	7%	2%	10%
Internet Index Fund	22%	–20%	30%

At the risk of restating the obvious, these assumptions may be reasonable based on historical performance of these securities; however, historical patterns may be useless for predicting future outcomes. Particularly for the Internet index, the available history is so short that we assume that historical rates of return (e.g., the 300+% return in 1999) cannot be sustained over a 25-year period. Therefore, the assumed rate of return on Internet stocks has been reduced here to 22%. Whether this assessment is too high or too low is simply unknowable in advance. We have assumed that as the rate of return increases, so does the risk of loss. Hence the Internet index is assumed to have the potential under a pessimistic scenario to destroy wealth at the rate of 20% per year.

Table 13-3 summarizes the results of various portfolio-allocation decisions for different types of investors under three scenarios. This analysis assumes that the middle-income investor invests $5,000 per year in various portfolios, that the upper-income investor invests $10,000 per year in the various portfolios, and that the affluent investor invests $40,000 in the various portfolios.

TABLE 13-3. MIDDLE-INCOME, UPPER-INCOME, AND AFFLUENT INVESTOR: 2025 VALUES FOR THREE PORTFOLIOS UNDER BASE-CASE, PESSIMISTIC, AND OPTIMISTIC SCENARIOS

Scenarios	Middle-Income 2025 Portfolio	Upper-Income 2025 Portfolio	Affluent 2025 Portfolio
Target Portfolio Size	$1,000,000	$3,000,000	$10,000,000
Portfolio 1			
Base Case	$890,093	$1,780,186	$8,158,164
Pessimistic	$198,005	$396,011	$2,579,438
Optimistic	$1,544,020	$3,088,039	$12,732,814
Portfolio 2			
Base Case	$1,910,700	$3,821,399	$24,240,510
Pessimistic	96,318	$192,635	$955,005
Optimistic	$4,386,708	$8,773,416	$57,194,910
Portfolio 3			
Base Case	$670,976	$1,341,935	$8,081,654
Pessimistic	$234,402	$468,805	$2,593,649
Optimistic	$1,031,690	$2,063,381	$12,732,814

Portfolio 1 Allocation = (S&P 500—60%, Money Market—30%, Internet Index—10%)

Portfolio 2 Allocation = (S&P 500—50%, Money Market—20%, Internet Index—30%)

Portfolio 3 Allocation = (S&P 500—55%, Money Market—40%, Internet Index—5%)

Source: Peter S. Cohan & Associates Analysis

The analysis summarized in the table provides a murky answer to the question of whether an investor should buy Internet stocks. The simple answer certainly depends on whether you agree with the assumptions used in this analysis. Even if this is so, there is still not a clear answer, because under some scenarios and with some portfolios, the different groups of investors may exceed their goal portfolio amounts or fall short of these amounts—even destroying a substantial amount of the value of their original portfolios.

If we assumed that each of the three scenarios was equally likely to occur, then the answer would be that all investors should buy Internet stocks in the proportion detailed in portfolio 2. In this case, middle-income, upper-income, and affluent investors would all invest 50% in the S&P 500, 20% in a money market fund, and 20% in an Internet index fund. Each year, these investors would plow 10% of their pretax income into these three investment categories in the aforementioned proportions.

If each of the scenarios proved equally likely, then the expected value of each type of portfolio would exceed their target value. For example, the middle-income investor would end up with a portfolio worth $2,131,242, exceeding the target amount by over $1 million; the upper-income investor would have a nest egg worth $4,262,484, exceeding the target amount by over $1.2 million, and the affluent investor would have a net worth of $27,463,475, exceeding the target by $17 million.

If the pessimistic scenario were twice as likely as the other two, then the expected value of each type of investors' portfolio would not exceed the target value by nearly as much. For example, the middle-income investor would end up with a portfolio worth $1,574,352, exceeding the target amount by over $500,000; the upper-income investor would have a nest egg worth $3,484,188, exceeding the target amount by under $500,000; and the affluent investor would have a net worth of $21,029,823, exceeding the target amount by a still-hefty $11 million.

Simply put, whether or not you should invest in Internet stocks at all depends on a number of factors. If you believe that Internet stocks are likely to outperform the S&P 500 with much more downside risk, as we have assumed here, then our analysis suggests that you should invest in a portfolio of 30% Internet stocks, 50% S&P 500 stocks, and 20% money market funds. Based on the assumptions used here, it is possible that investors with different income levels and investment goals would be able to exceed those goals under a variety of scenarios.

Unfortunately, there is no way of knowing whether any of these assumptions is likely to occur. However, conducting the kinds of scenario analyses we have done here can alert investors to the most important assumptions. It is up to each investor to conduct such an analysis for themselves to determine what kinds of risks and returns they are comfortable with. If an investor concludes that he or she is comfortable with the assumptions about future risk and return in Internet stocks in general, then it makes sense to consider how best to invest in Internet stocks.

The question of whether or not to invest in Internet stocks then comes down to one of whether or not you believe that they offer the potential for greater return at manageable levels of risk. I believe that the qualitative advantages outweigh the disadvantages of investing in Internet stocks; however, investors must draw their own conclusion. I hope that this book has provided you with the tools to do just that.

In any case, here is my list of advantages of investing in Internet stocks:

- Internet stocks have offered very high rates of appreciation in the stock market since the first Internet-related stocks entered the public securities markets in 1995. As we have noted repeatedly throughout this book, the 1999 rate of return for the nine Internet segments exceeded 300%.
- Internet stocks represent a way to invest in the fastest-growing product/service market in the world, and the limited penetration of that market—as we noted in Chapter 2—suggests that the potential for growth has yet to be fully realized.
- Some Internet sectors offer clear profit potential for companies. Given the rapid growth rate of the markets that these sectors serve, it is likely that these companies will be able to generate exceptionally high levels of profit and profit growth.
- New Internet-related businesses, such as wireless Internet access and wireless e-commerce, are likely to create significant new opportunities for rapid revenue and profit growth for investors who can make astute decisions regarding the companies that are best positioned to tap into this growth.

Here are some of the disadvantages of investing in Internet stocks:

- Internet stocks can go down sharply and quickly. For example, in the springs of 1999 and 2000, Internet stocks experienced declines in excess of 40% within the course of two weeks. In April 2000, for example, some e-commerce companies saw their stock prices decline in excess of 70%, and many of these companies are unlikely ever to return to their all-time highs.

- Internet business is harshly Darwinian. This means that many new firms may be started to take advantage of a perceived business opportunity, but only a few firms will survive. It is important that investors in Internet stocks develop the skills to pick the winners because the winners are likely to prosper in both the product and the securities markets.

- Internet stocks are subject to rapid movements based on the behavior of momentum investors. Because Internet companies typically take public a fairly small proportion of their total shares, the relatively few outstanding shares of an Internet company are likely to be more volatile. This higher level of volatility means that the investment decisions of a small number of shareholders can lead to unusually wide swings in a stock.

SHOULD YOU INVEST IN AN INTERNET STOCK INDEX OR IN SPECIFIC INTERNET SECTORS?

If you are still reading, it is likely that you are leaning toward agreeing with my conclusion that the advantages of investing in Internet stocks outweigh the disadvantages. If you are interested in investing in Internet stocks, I strongly recommend avoiding the Internet stock index and focusing instead on the most attractive sectors based on our reviews in Chapters 4 through 12. The reason for this recommendation is that our analyses have highlighted specific sectors of the Internet economy that are likely to perform poorly in the product and stock markets and other sectors that are likely to outperform their peers. I believe that the returns are likely to be greater than average for investors who stay away from the weaker Internet sectors and focus on the most attractive ones.

As we noted in Chapter 3, there have been major differences in the stock market performance of the different Internet sectors. Figure 13-1 summarizes the performance of these different sectors in 1999 and 2000. The companies included in the nine Internet stock sectors analyzed here increased 339% in value in 1999 but declined 67% from their 1999 year-end levels in 2000.

There were significant differences among the sectors in terms of stock market performance during the period of rising returns (1999) and the period of declining returns (2000). An analysis of the companies in the various sectors suggests that the strongest firms did the best on the upside and retained their value the most during a period of decline. In fact, some of the winners in the strongest segments actually saw an acceleration in their stock prices during the first half of 2000, as investors sought a place to invest the cash they had raised by selling shares of some of the weaker industry participants.

Many of the firms whose stock prices increased after the crash were analyzed in Chapters 4 through 12. These postcrash winners include Web security firms Check Point Software (+169%) and ISS Group (+10%), an infrastructure firms, Juniper Networks (+122%), Web content provider Forrester Research (+45%), and ISP IDT Corporation (+8%).

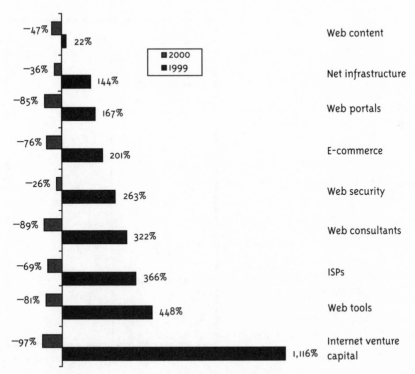

Figure 13-1. Stock Price Performance of Nine Internet Sectors, 12/31/98–12/31/99 and 12/31/99–12/29/00

Source: MSN MoneyCentral; Peter S. Cohan & Associates Analysis

The point of this analysis of Internet sectors in up and down markets is to suggest that investors can achieve stock market returns that are much better than the Internet index average by selecting specific sectors in which to invest. The question for investors is to identify the specific sectors that are likely to generate the highest returns for investors at acceptable levels of risk. Figure 13-2 provides a framework that investors can use to conduct such analysis. The basic notion is that if investors analyze the performance of a sector in up- and down-markets, they will find the ideal sectors in which to invest are the ones that appreciate more than the average in up markets and drop less than the average in down markets. Depending on risk preference, investors might find that the second-most attractive sectors in which to invest are the ones that appreciate less than the average in up markets but drop less than the average in down markets.

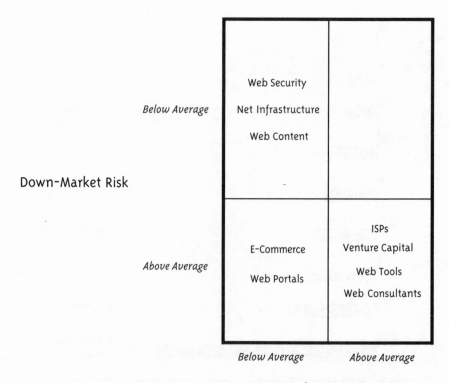

Figure 13-2. Internet Sector Stock Market Performance in Up and Down Markets

While it would be a mistake to rely solely on the analysis in Figure 13-2, because past performance is not a guarantee of future outcomes, this analysis suggests some priorities for investors. For example, the analysis makes it clear that investing in e-commerce stocks is a mistake in both rising and declining markets. The irony of this finding is that e-commerce and Web portal stocks have received a disproportionate share of attention in the business press since the mid-1990s. However, the recent performance of e-commerce and Web portal stocks suggests that they should be avoided. This finding actually seems consistent with the analysis we conducted in Chapters 7 and 8, which suggested that it is likely to be difficult for e-commerce and Web portal firms to sustain profits, at least for the foreseeable future.

Interestingly, Figure 13-2 suggests that there are no Internet sectors that are safe in both up and down markets. In 1999, the ISPs that we analyzed lost money; however, they did well in the stock market. For example, the ISP sector did extremely well in 1999 because of Metricom and Exodus. In 2000, Metricom's stock lost 87% of its value from the end of 1999, and Exodus's stock lost 55%. The biggest winners in the stock market for ISPs were two firms that agreed to be acquired in 2000, Verio and Concentric Networks. This kind of stock market behavior could continue; however, my preference would be to stick with sectors where economic bargaining power is strong—not a characteristic of the average ISP.

Depending on risk preferences, an investor may wish to take a harder look at the Internet sectors in the upper-left-hand corner of Figure 13-2. Of the three sectors, Internet infrastructure appears to be of greatest interest. Although the average stock in the sector increased a relatively paltry 144% in 1999, Internet infrastructure stocks lost a relatively modest 39% in 2000. It is likely that selected Internet infrastructure stocks will do well in 2001 and beyond.

The Web security sector could also be worth looking at, as we noted in Chapter 9, because it contains a mixture of very weak players and some companies that could become stronger in the future. The Web content sector appears to be worth avoiding insofar as it has numerous weak players and does not appear to have a strong claim on gaining economic bargaining power. Furthermore, its relatively weak 22% increase in value in 1999 does not make 2000's lower-than-average 47% decline in value particularly attractive as a store of value. Investors would probably be better off investing in a money market fund than in a Web content firm.

The four sectors in the lower-right-hand box of Figure 13-2 are likely to be of greater interest to investors who are willing to take on more risk to get higher returns. Of these four Internet sectors, Web tools is likely to offer the best performance to the discriminating investor. As we discussed in Chapter 12, the Web tools sector includes an array of firms with widely varying levels of economic bargaining power. Investors are likely to enjoy higher returns by focusing on the firms with greater economic bargaining power and shun those in a weaker position.

The other three sectors in this lower-right-hand box do not offer much promise for investors in the intermediate term. The Web consulting firms were hit hard during the 2000 Internet stock correction, although the firms with greater economic bargaining power, based on their ability to deliver on a closed-loop solution that generates high client satisfaction, held up better. Nevertheless, investing in this sector could be a bit of an uphill battle for a while. Investing in the Internet venture capital sector is likely to be extremely volatile because it remains very difficult for investors to assess the value of the firms' portfolios. Finally, Web portals appears likely to continue to consolidate. However, in the absence of an unexpected acceleration in the profit growth of Yahoo, the pickings in the Web portal sector may be slim.

To recap, Internet investors should certainly avoid buying an Internet stock index if they are unwilling to conduct some analysis to dig out the sectors that are most likely to meet their specific preferences for risk and return. Recent performance of the sectors suggests that investors may wish to avoid e-commerce stocks. As for sectors to focus on, we have identified ISPs, Internet infrastructure, and Web tools as offering the greatest upside potential and the most downside protection.

CRITERIA FOR INVESTING IN INTERNET COMPANIES

Investors seeking to pick specific stocks within these sectors can use the IID framework we introduced in Chapter 1 and have been developing in Chapters 4 through 12. At the risk of repeating what should by now be second nature to readers, I will emphasize the elements of the IID framework that I believe are likely to be of most use to investors. In this section on monitoring the portfolio of Internet stocks, we will discuss some specific sources of information that investors can use to assess changes in the key IID indicators.

The IID framework is intended to be thought of as a giant filter. The universe of Internet stocks goes into the top of the filter and a small number emerges at the bottom. These filters are finely tuned to protect investors from investing in the latest fads. For example, these filters would have kept investors from buying B2C e-commerce stocks because of their very low economic bargaining power.

Let's review the six parts of the IID filter:

1. **Does the industry have economic bargaining power?** This test is first in line because it knocks out so many of the Internet industry sectors. As we noted earlier, Internet infrastructure and Internet venture capital are among the sectors that have economic bargaining power. They control a scarce resource that is in demand by powerful decision-makers. Economic bargaining power can be measured by the extent to which a firm can charge a price in excess of its costs, as reflected in the firm's net margin. Firms like Cisco Systems and Check Point Software have high net margins because they have economic bargaining power.

2. **Does the company offer customers a closed-loop solution?** This test is important because firms that offer customers closed-loop solutions—as measured by industry analysts and interviews with customers themselves—tend to have higher market shares and are thus in a stronger position to create virtuous cycles that enable them to grow faster and more profitably than their peers. AOL and Macromedia are forging such closed-loop solutions, and they are experiencing accelerated revenue growth even as their peers are slowing down and merging with bigger partners.

3. **Does the firm's management have integrity, and does it adapt effectively to change?** If a firm passes the first two tests, investors need to know whether the firm can continue to win as the competitive environment changes. Such change is inevitable and unusually rapid in Internet business. As a result, investors need to know whether a firm's management can live up to its commitments to stakeholders and whether it can deliver consistent earnings growth in a rapidly changing environment. As we saw in Chapters 4 through 12, a number of firms that proved to

be poor investments were not able to pass this test. Examples include Excite@Home, Network Associates, and Open Market.

4. **Does the company create a compelling brand family?** Many investors do not understand the technical details of an Internet company's business. As a result, these investors rely on signals of value. These signals of value spring from the collection of customers, partners, and investors that the Internet company's management team is able to assemble. If these customers, partners, and investors are perceived as top-tier, then investors are likely to want to join the party. If not, investors are likely to be less interested. During our foray into the nine Internet business sectors, we identified several companies, such as Juniper Networks, Ariba, and BroadVision, which had done an excellent job of assembling a compelling brand family. In many cases, this brand family helped to bolster the firms' stock prices.

5. **Does the company make effective use of its financial resources?** As we have noted throughout this book, investors began to place an increasing emphasis on financial effectiveness following the April 2000 crash in Internet stocks. This crash made investors take a much sharper look at whether firms had sufficient cash flow to cover their cash-burn rate. Given the desiccated environment for financing, the management of many B2C e-commerce firms has only been able to extend the time to tank-empty by cutting back on spending. As we noted earlier in the book, firms that found themselves too close to the tank-empty position found themselves in single-digit stock price territory—a difficult hole out of which to climb. We looked at firms like CDNOW, which were in that hole and were ultimately acquired. We also found companies like Forrester Research, which exhibited remarkably disciplined financial management even as its peers were spending as if they would never need to concern themselves with running out of money.

6. **Is the company's stock market valuation relatively inexpensive?** As we saw in analyzing the market valuations of Internet stocks, the key metrics are the price-to-sales ratio and price-to-sales divided by revenue growth rate. Comparing firms within a specific Internet sector yields insights into which firms within a sector may be worth evaluating to assess whether or not they are undervalued and hence worth considering as an investment. For example, in the ISP area we found that IDT was valued much more cheaply than its peers. As we noted in

our discussion, this relative valuation should be weighed in conjunction with the other five factors to determine whether this relative valuation is a temporary mistake in the market, which astute investors could exploit. The example we cited earlier of Check Point Software's August 1998 price of $5.50 a share was clearly such an opportunity.

Whether there are individual companies that pass these six tests is a question for investors to answer for themselves. During the course of analyzing the nine Internet business segments as of this writing, we identified a number of interesting opportunities. Whether these opportunities are still interesting as you read this depends on your diligence in analyzing them and other potential investment opportunities using the tools illustrated in this book. In general, my recommendation is simple to articulate and hard to achieve: Before you invest your money, invest your time.

TIMING OF INTERNET STOCK PURCHASES AND SALES

Timing of Internet stock purchases and sales is a question that investors in any stock must address. One school of thought is to try to identify a temporary market bottom and purchase a significant number of shares near that bottom. Another approach is to invest a fixed dollar amount on a regular basis in stocks that pass the six tests just reviewed. Under this fixed-dollar-amount approach, investors regularly purchase more shares when the stock's price is lower and fewer shares when the price per share rises.

As for when to sell, one approach is to set a target rate of return and then to sell a specific percentage of ones' holdings should that target be reached. If the investor continues to believe that the company's potential for further stock price increases is great, then the investor should hold a proportion of the shares. In this case, the investor should set higher price targets, and if the shares hit that new price target, the investor should then reevaluate whether the company's strategic positioning remains sound.

Let's evaluate the first buying strategy—picking a temporary market bottom to load up on shares. This approach has one obvious advantage: If the investor guesses right, there can be a tremendous increase in the stock price in a relatively short period of time as investor sentiment improves. The approach also has several disadvantages. First, it is impossible to know whether the bottom is temporary or semitemporary. In other words, investors may pick what they think is a bottom only to real-ize after the investment has been made that the market has another 20% further down to go. Conversely, this approach may cause investors to delay investing in a good company at a low price because they think the market has further to fall.

One question that can at least be analyzed based on past history is the notion that Internet stocks have a certain seasonal pattern. In other words, it is possible to look at historical price performance of the Internet indices and identify whether there are periods when the index seems to drop and other periods when it seems to rise. As Figure 13-3 (repeated from Chap-ter 3) indicates, there seem to have been relative dips in the month of April. In 2000, this April dip was huge, many Internet stocks dropping 40% to 60% in the course of a week.

As it turned out, this dip was an excellent short-term buying opportu-nity for selected stocks. For example, Macromedia stock dropped from about $95 in March to $42 in mid-April. By mid-June, Macromedia was trading at $114 a share. Thus, if an investor had mustered the courage to buy Macromedia in mid-April at $42 and sold in mid-June, he or she would have almost tripled their money in two months. A similar dip took place in 1999 for many Internet stocks. Whether this pattern will continue is hard to know.

Such drops can be extremely valuable opportunities for astute investors. First, investors must be prepared to act when these drops occur. By being prepared, I mean knowing how the IID analysis applies to all the stocks in the Internet sector. In the case of Macromedia, investors who had conducted the IID analysis would have realized that Macromedia's underlying business was just as strong when it was trading at $100 a share in March 2000 as it was when its stock had dropped to $42 a month later.

What happened in April 2000 is that investors sold everything in a panic and asked questions later. Many of the stocks that investors sold were e-commerce companies that lacked economic bargaining leverage. As of this writing, many of these company's stock prices are trading at 60% or more below their precrash levels and may never recover. By contrast, firms that passed the six tests of the IID analysis were magnets for investor money, climbing rapidly back as investors recognized that the crash had created some excellent buying opportunities. The courageous act on the part of investors would have been to purchase shares of companies like Macromedia at $42, not waiting to see if they might drop still further.

Is there an April effect in which Internet stocks tend to drop in April

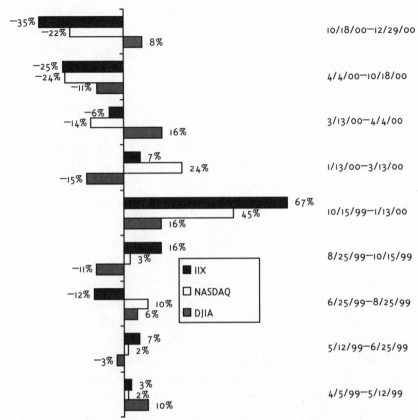

Figure 13-3. Performance of Dow Jones Industrials, NASDAQ, and Internet Index
(IIX) in Up- and Down-Market Periods, April 1999—December 2000
Source: PCQuote, Peter S. Cohan & Associates Analysis

every year and then rise later in the year? Anecdotal evidence suggests that there might be such an effect. On the NASDAQ, the seven Mondays preceding 2000's tax day were down. This trend started in late February, when the early tax-filers were preparing their 1040 forms, and continued through April 10, as last-minute filers rushed to complete their taxes.

Mondays in the weeks before taxes are due tend to have sell-offs, since Americans who spend the weekends filling in their 1040 forms often realize that they owe the government a big payment. Many Americans use their brokerage accounts as their savings accounts, and many of them do not have enough money in their bank accounts to pay the tax bill, so they sell some stocks to raise cash.

In 1999, three of the five Mondays before tax day saw NASDAQ declines. However, after tax day passed, the following three Mondays were all up. While it is unclear why the April theory affects only the NASDAQ index, it could be that many momentum traders—investors who get in and out of positions relatively quickly, thus incurring bigger capital gains taxes—invest only in technology stocks and thus only have technology stocks to sell when their tax bill comes due.

If this April effect persists, it suggests an obvious opportunity for investors in Internet stocks. Investors should identify Internet stocks that pass the six IID tests and monitor them closely as tax day approaches. If the April effect takes place and the prices of the identified Internet stocks drop, investors should consider buying shares of these stocks after tax day. If history is any guide, this should prove to be an excellent buying opportunity. On the other hand, if a sufficient number of investors begin to follow this trading strategy, it is likely that the profitability of this strategy could decline.

Before concluding this discussion of the timing of purchase and sale decisions, let's repeat the key lessons. The first key lesson for investors is to invest time in picking individual Internet stocks that are likely to do better than their peers in sectors with economic bargaining power. Second, seasonal behavior in markets has been noted in the past, but it is not clear whether that behavior is a reliable predictor of future market behavior. Nevertheless, if seasonal phenomena such as the April effect persist, investors should consider using such effects to time purchases. Third, investors should consider whether to make one big investment at what they perceive as a market bottom or to dollar-cost-average their purchases, the strategy we mentioned earlier.

Finally, investors should set price targets at which they will sell their stocks. These price targets should be both high and low. If a stock reaches the high target and the company still looks as if it will continue to perform well—because of its attractive market, compelling strategy, outstanding management, etc.—then investors may consider holding part of their position rather than selling everything outright, then setting a higher price target for the remainder. If a stock declines below its floor price (e.g., 7% below the price at which the investor purchased the shares), then the investor should make a similar assessment of the company's fundamentals, using the six IID tests. If the company's fundamentals have eroded, the investor should strongly consider selling the entire position in the stock. If the company's fundamentals are still strong, then the investor may consider selling a small portion of the shares or even buying more shares at the lower price.

MONITORING THE PORTFOLIO OF INTERNET STOCKS

Monitoring a portfolio of Internet stocks involves tracking macroeconomic factors as well as the IID indicators for each stock in the portfolio. First, let's focus on identifying the key macroeconomic indicators driving Internet stock prices and how best to monitor them. Then we will describe how to monitor the IID indicators.

Macroeconomic Indicators

Internet stock prices in general are driven by a number of macroeconomic factors. These factors are external forces, including traditional factors such as interest rates, tax policies, and fiscal policies. The key macroeconomic factors—and the means of monitoring them—are described below:

- **Interest rate policy.** Interest rates affect the price levels of all stocks, including Internet stocks. If the Federal Reserve is raising interest rates and investors expect those rate increases to continue, then it is likely that most stock prices will decline. If, however, there are signs that the economy is slowing down, investors may begin to suspect that the Federal Reserve is less likely to raise interest rates in its next meeting. As soon as investors suspect that a series of Federal Reserve rate increases is likely to end, they may begin buying shares of technology stocks, particularly those Internet stocks that are not dependent on borrowing money to finance their operations. This caveat regarding borrowing is important because, as interest rates rise, the cost of financing rises as well. If a company borrows money, it is likely to be hurt by the rise in interest rates. Since many Internet companies do not borrow money, their growth is not likely to be affected. Conversely, if the economy is in a recession and the Fed is cutting interest rates, this can often be a great time to purchase the shares of some of the high-quality Internet stocks whose shares may have been beaten down in a general flight out of technology stocks.
- **Internet financing trends.** Investors should recognize that an important factor driving the value of Internet stocks is the changes in supply and demand for the shares. Internet companies are funded largely by

venture capital firms. If the level of capital being raised by venture funds is growing and corporate and angel investors are pouring money into Internet companies, then it is possible that the number of potential shares of Internet companies will rise in the future. This supply of shares ultimately depends on the extent to which there is a market for IPOs—i.e., whether or not the demand for shares in these ventures is strong. Typically, the supply and demand for Internet company shares goes in waves. The wave typically starts off with a small number of Internet companies getting financing during a period when venture capital funds have little money—often during a general economic recession. During this phase, the quality of the companies receiving financing is generally higher. As the general economy emerges from a recession, so does the IPO market awake from its slumber. At this point, the highest-quality Internet companies go public and often do very well in an IPO market very hungry for new deals. This begins the second phase. Here a significant amount of new money becomes available for financing Internet startups. The venture firms see their coffers swell as they take profits from their IPOs. Corporations and pension funds seek to invest more money in these successful venture firms. Angel investors begin to invest, as do corporate venture funds. During this second phase, the amount of money available to be invested is so great that the quality of the companies financed tends to decline, and the level of consideration for each portfolio company diminishes as well. The phase-one success of the IPO market almost ensures that the appetite for new IPOs will be tremendous during phase two. Phase-two IPOs tend to do very well initially, but then a shakeout occurs in which investors dump shares in the lower-quality IPOs and hold the shares of the top firms. In the third phase, the IPO market dries up, and the window for financing Internet startups closes perceptibly. This third phase is characterized by layoffs, mergers, and a general pullback that leads to more investment discipline—returning after a few years to phase one. Investors should monitor the phases of the market for financing IPOs. It is generally best to invest during the bleakest period, toward the middle of phase three, and to avoid investing just before the peak of phase two.

- **Government.** The federal government can have a significant impact on the value of Internet stocks. For example, government has the power to tax e-commerce. While it has held off on such taxation as of this writing, it is entirely possible that e-commerce taxes could be imposed in the future once the majority of people are dependent on e-commerce. In short, government does not yet want to risk terminating a potentially lucrative source of tax revenues—e-commerce—by imposing taxes too early in its development. Similarly, government has the potential to facilitate security measures for e-commerce, to regulate how data about Web surfers' online habits are used, and to determine the ease or difficulty of enabling non-U.S. technology workers to put their talents to use for Internet businesses. Investors should be tracking the role of government in the Internet business and assessing potential changes in this role as well as the way that these changes might alter the prices of Internet stocks.

For investors, the sources of information for tracking these macroeconomic factors are very good. For example, MSNBC, the *Wall Street Journal, Barron's*, and many other magazines and newspapers do a good job of reporting how these macroeconomic factors are changing. However, it is often up to the investor to assess how these factors are likely to influence the behavior of Internet stocks.

To summarize, here are some of the changes in macroeconomic factors that should make investors lean toward selling Internet stocks:

- Inflationary economic indicators, such as higher Consumer Price Index (CPI) and Producer Price Index (PPI) levels or lower unemployment rates, that suggest the Federal Reserve could begin raising interest rates
- An increase in the number of IPOs, which suggests that the quality of IPOs is declining and that supply could exceed demand, thus lowering values
- A dramatic rise in the average first-day IPO price gain, followed by subsequent price collapses within weeks of the IPO—another phenomenon suggesting that the quality of supply is declining
- An erosion in the strength of the income statements and balance sheets of companies being taken public—another phenomenon suggesting that the quality of supply is declining
- Significant discussion about the IPO market and the level of stock prices among people who work at low-wage jobs, which suggests that many people purchasing IPO shares do not understand the fundamen-

tals of the companies whose shares they are buying and that they may be more likely to sell indiscriminately in the event of a market decline

- General media coverage of Internet executives and venture capitalists as modern-day heroes, which suggests that the media will be presenting an overly glamorous picture of Internet business to the general public and will not educate the public about the risks of Internet business

- Government passing legislation that impedes Internet industry development—taxing e-commerce, limiting the number of workers from outside the U.S. who can work in U.S.-based Internet companies, changing accounting policies to require companies to account for mergers as purchases and thus creating huge goodwill charges for acquirers, or forcing Internet companies to include stock options to employees as a compensation expense—making it harder for Internet businesses to develop profitably

Now let's turn to some of the factors that Internet stock investors should monitor that are specific to the value of the individual companies in which they have invested. Investors should track these factors for each of the Internet sectors in which they own stocks or are considering buying stocks. Here are six factors that investors should monitor because changes in them might suggest buying or selling shares:

1. **Industry economic bargaining power.** Investors should track changes in profit margins for publicly traded companies in the sector each quarter. If the general level of profit margins is negative and declining, this suggests that the sector lacks economic bargaining power, and investors should avoid the stocks in the sector. If there are changes in the drivers of economic bargaining power, these changes could find their way into the net margins over time. For example, if demand for a sector's product is growing even as the number of suppliers is dropping—because of mergers or bankruptcies—this could indicate that the profit margins of the remaining participants could increase in the future. Such trends could suggest industry economic bargaining power would improve and could signal a potential buying opportunity.

2. **Closed-loop solutions.** Investors should monitor customer feedback and analysts' assessments of competitors' products and services.

Technology trade publications are generally available online and often include interviews with customers who have used products from different vendors within a sector. Analysts often rate the quality of these products and services, and the rankings are particularly useful when correlated with customer feedback. Investors should also monitor changes in relative market share and relative levels of revenue and revenue growth. As a general rule, firms that get very positive feedback from customers and analysts and enjoy leading relative market share and rapid revenue growth tend to be the ones that are offering closed-loop solutions. Companies that offer closed-loop solutions tend to sustain market leadership, finding themselves in a stronger position to capitalize on industry consolidation. Investors should buy shares of the firms offering closed-loop solutions and sell the rest—unless they have reason to believe that lagging firms will be sold to leaders at above-market prices.

3. **Management integrity and adaptability.** Investors should monitor changes in legal proceedings and accounting policies to assess changes in management integrity. As a general rule, problems with management integrity do not go away once they emerge. These problems are sometimes an early-warning indicator of problems in the future. Unfortunately for investors, by the time such problems are printed in financial statements, it is too late to get out before a stock price decline. Investors should also look for evidence of adaptability among the firms in an Internet sector. Consistent earnings growth is a rare phenomenon in Internet business; however, if investors can find companies that exhibit such consistency, they should seriously consider purchasing their stock. Investors should also monitor more qualitative indicators of management adaptability, such as a history of well-integrated acquisitions or the ability to build new businesses internally that generate significant revenues. Firms that exhibit high integrity and adaptability are likely to make good investments. Firms that fail in these categories are often ones for investors to avoid.

4. **Brand family.** Investors should assess the quality of a firms' customers, partners, and investors and track how the brand families change each quarter. Investors should buy stock in firms that improve the quality of their brand families over time. If the quality of customers and, in particular, partners is improving over time, this suggests that management is creating a virtuous cycle that is likely to

garner the company the lion's share of available profit in the market. If the quality of customers and partners is declining, this suggests that selling the stock might make sense.

5. **Financial effectiveness.** Investors should also monitor firms' financial effectiveness over time. Firms with growing sales productivity and level or tightening SG&A-to-sales ratios are exhibiting a concern for shareholder return that should signal upside in the stock. This trend must be monitored in conjunction with other factors because investors would not want to buy shares in a company that is starving key activities for the sake of increasing short-term profits. Investors must be cognizant of which companies are striking an appropriate balance for the sector between frugality and profligacy. Investors should consider purchasing shares in the firms exhibiting balance and consider selling the shares of the others.

6. **Relative stock market valuation.** Investors should also track changes in the price-to-sales ratio and price-to-sales divided by revenue growth rates. If a firm that performs relatively strongly on the other five indicators has a low market valuation, this could be a great buying opportunity. Investors should also monitor changes in the criteria used to monitor valuation. For example, if various Internet sectors begin to generate significant profits, investors will begin to use price/earnings ratios, not price/sales ratios. Conversely, if more companies begin to run out of cash, investors may begin adopting time-to-tank-empty as the key indicator of Internet company value.

Fortunately for investors, the Internet provides a wealth of resources for monitoring these six factors. Investors should consider setting up personal portfolios on Yahoo Finance with the individual Internet stocks in their portfolio. Over time they can monitor news influencing the individual stocks and the industry from the Yahoo Finance page. To monitor general industry developments, investors should consider reading the online versions of *Upside, Red Herring, The Industry Standard, Business Week, Fortune, Business 2.0, CNET, ZDNet*, and *TechWeb*. Finally, Microsoft's MoneyCentral is an excellent source of integrated information for analyzing individual companies. I encourage investors to spend time to know what is driving the value of their Internet stocks. These sources are useful for gaining this insight.

CONCLUSION

Internet stocks can generate huge returns for investors—and significant losses. Despite all the hoopla about Internet stocks, there is a systematic method that the average investor can use to find the great long-term performers and avoid the rest. This method is based on indicators that are measurable. Profiting from this method demands discipline and an ability to resist the siren song of the media. My experience suggests that the rewards of persistently following this discipline are well worth the investment.

REFERENCES

Bary, A. "Tortoise and Hare." *Barron's*, March 13, 2000. [http://interactive.wsj.com/articles/SB95273956295244461.htm]

Byron, C. "Defibrillating the Dot-Coms." MSNBC, February 15, 2000. [http://www.msnbc.com/news/370477.asp?0m=-14T]

Donlan, T. "Virtual Reality." *Barron's*, February 14, 2000. [http://interactive.wsj.com/articles/SB950313512680235752.htm]

"East vs. West." *Delaney Report,* July 19, 1999, p. 3.

Evans, J. "The Fox Network." *The Industry Standard,* April 24, 2000. [http://www.thestandard.com/article/display/0,1151,14288,00.html]

Ewing, T. "Burnt Offerings? Street Debuts Are Fizzling After Pop." *Wall Street Journal,* April 26, 2000, p. C1.

Landry, J. "Econets Suffer from Fatal Flaws." *Red Herring*, November 27, 2000. [http://www.redherring.com/vc/2000/1127/vc=divine112700.html]

Masud, S. "Transforming the Network Core." *Telecommunications,* February 2000, p. 36.

Meyer, L. "Prepare for a Parade of Poor Earnings." *Red Herring,* January 10, 2001. [http://www.redherring.com/investor/2001/0110/inv-earnings011001.html]

Schwartz, N. "Meet the New Market Makers." *Fortune,* February 21, 2000. [http://www.pathfinder.com/fortune/2000/02/21/psy.html]

"Telecommunications Equipment Market 'Undergoing Tremendous Fundamental Change,' New Sutro Report Points Out." *Cambridge Telecom Report,* September 6, 1999.

"Tiger Management to Close Down." Reuters, March 30, 2000. [http://www.msnbc.com/news/388494.asp]

Webber, A. "New Math for a New Economy." *Fast Company,* Jan./Feb. 2000. [http://www.fastcompany.com/online/31/lev.html]

Willoughby, J. "Burning Up." *Barron's,* March 20, 2000. [http://interactive.wsj.com/archive/retrieve.cgi?id=SB9533335580704470544.djm]

INDEX